COMBUSTION CALCULATIONS

COMBUSTION CALCULATIONS
Theory, worked examples and problems

E. M. Goodger
School of Mechanical Engineering,
Cranfield Institute of Technology

First published 1977 by
THE MACMILLAN PRESS LTD
London and Basingstoke
Associated companies in New York Dublin
Melbourne Johannesburg and Madras

ISBN 0 333 21801 9

Printed and bound in Great Britain
by Unwin Brothers Limited,
The Gresham Press, Old Woking, Surrey

iv

CONTENTS

PREFACE

Examination syllabuses concerned with combustion reactions and their products tend to draw on material appearing in texts on chemistry, physics, thermodynamics, fuel science and the few publications devoted to combustion itself. Furthermore, some aspects of the subject tend to be handled in a 'case law' manner by depending solely on worked examples to illustrate the techniques of solution, whereas a prior general statement of the relevant theory permits much wider application.

The objectives of this present book, therefore, are threefold.

(1) To assemble in one small volume the key relationships of stoichiometry, thermochemistry and kinetics that apply to the calculation of combustion quantities.

(2) To unify these relationships by starting with a general combustion equation, in simple terms of molecular products only, and adapting this systematically to each aspect in turn.

(3) To provide solutions in general terms before illustrating their use in examples typical of examination questions met in combustion and allied fields.

The motivation for a work of this kind stemmed from experience with examination marking at H.N.C., H.N.D. and first- and higher-degree levels in various branches of engineering and fuel technology, where it has become apparent that questions of this type present difficulties to some candidates. This approach can also serve in professional practice as a reminder of the basic principles involved before turning to the more sophisticated instrumentation and computational techniques available in industry and research.

The book opens with a brief review of the main features of atmospheric air and conventional fuels, together with those of the most common combustion products and their methods of measurement. The following chapter deals with the determination of the proportions of reactants, and of the cooled, stable products of combustion. Chapter 5 is concerned with the determination of proportions of hot products that are still reacting but have reached some condition of dynamic equilibrium. Chapter 6 deals with the quantities of energy involved in combustion reactions, and chapter 7 with the temperatures reached under adiabatic, or some other specified, conditions. The efficiency of combustion is covered in chapter 8, whereas chapter 9 indicates the enhanced efficiency realised by electrochemical oxidation in the absence of substantial heat release. All key equations are numbered in sequence, and listed in a final summary. As a general rule, decimal fractions are used, but it has occasionally proved more convenient to include such a term as $(x + y/4)$ rather than $(x + 0.25y)$. Worked examples are included at the

ends of chapters, and a number of problems provided, with answers.
SI units are used throughout, and extracts from the wealth of thermochemical data published in kcal/mol have been converted by the relationship

$$(X) \text{ kcal/mol} = 4.184(X) \text{ kJ/mol}$$

Acknowledgement is made gratefully to colleagues and students of
the University of Newcastle, N.S.W., and the Cranfield Institute of
Technology, Bedfordshire, for much valued assistance in discussion
and feedback, to the National Bureau of Standards for helpful com-
ments, and particularly to Professor R. S. Fletcher of the Cranfield
School of Mechanical Engineering, for the facilities that made this
work possible.

Cranfield, 1976   E. M. GOODGER

UNITS

In any system of units, certain quantities are defined as basic to
the system, and all further quantities derived from them. If the
system is coherent, the products and quotients of any two or more
unit quantities themselves become the units of the derived quantities. Thus, in the case of Newton's second law, F = ma, 1 derived
unit of force is equal to

(1 unit of mass) × (1 unit of length)/(1 unit of time)^2

One of the most significant developments in this area is the adoption of a rationalised system of metric units known as SI (Système
International d'Unités), which is coherent, with a derived unit of
energy common to the mechanical, electrical and most other forms.
SI includes the following base units

length	metre (m)
mass	kilogram (kg)
time	second (s)
electric current	ampere (A)
thermodynamic temperature	kelvin (K)
amount of substance	mole (mol)

It should be noted that the kelvin is also used for temperature
intervals, and that the mole relates to what was formerly called
the 'gram-mole' and not the 'kilogram-mole' (kmol).

SI includes the following derived units

force	newton (N) = kg m/s^2
pressure	pascal (Pa) = N/m^2 = kg/m s^2
energy	joule (J) = N m = kg m^2/s^2
power	watt (W) = J/s = kg m^2/s^3

No change is made to any symbol to indicate the plural.

SI has been adopted by various industries (for example, *Recommended SI Units,* Institute of Petroleum, London, 1970), some of
which also use earlier metric units such as the litre and the bar,
which are not part of SI but are considered acceptable. Since the
adoption of SI is not yet worldwide, the following conversion fac-
tors and other metric relationships are given.

length	1 ft = 0.3048 m
volume	1 ft^3 = 0.028 32 m^3
	1 U.K. gal = 1.201 U.S. gal = 4.546 litre
mass	1 lb = 0.4536 kg
	1 tonne = 1000 kg
density	1 lb/ft^3 = 16.0185 kg/m^3

force	1 lbf = 4.4482 N
pressure	1 lbf/in.^{2} = 6.894 76 kPa
	1 mm Hg = 133.322 Pa
	1 atm = 101.325 kPa = 1.01325 bar
	1 bar = 100 kPa = 10^5 Pa
energy	1 Btu = 1.0551 kJ
	1 Chu = 1.8991 kJ
	1 kcal (international table) = 4.1868 kJ
	1 kcal (thermochemical) = 4.184 kJ
	1 kWh = 3.6 MJ
	1 hp h = 2.6845 MJ
	1 therm = 10^5 Btu = 105.51 MJ
specific energy	1 Btu/lb = 2.326 kJ/kg
specific energy capacity	1 Btu/lb ^{o}R = 1 Chu/lb K = 4.1868 kJ/kg K
volumetric energy	1 Btu/ft^3 = 0.0373 kJ/litre (or MJ/m^3)
	1 Btu/U.K. gal = 0.232 kJ/litre (or MJ/m^3)
power	1 hp = 745.7 W

NOTATION

When two symbols are given for one item, upper case represents an extensive property (dependent on mass) and lower case a specific property (per unit mass). When one symbol is used for more than one item, the particular meaning in any instance will be apparent from the context.

A   air mass

A, a   non-flow availability function

$\dot{A}$ air mass flow rate

A/F   air/fuel ratio by mass or by volume

B, b   steady-flow availability function

C   molar heat capacity

c   specific heat capacity

C/H   carbon/hydrogen mass ratio

CV   calorific value

E   electrical potential

e   charge/electron

F   fuel mass

$\dot{F}$ fuel mass flow rate

G, g   Gibbs free-energy function

(g)   gas

GCV   gross calorific value

(gr)   graphite

H, h   enthalpy

i   electrical current

ip   indicated power

K   equilibrium constant

k   rate constant

(l)	liquid
M	number of moles of oxygen/gram of fuel
m	number of moles of oxygen/mole of fuel
$\dot{m}$	mass consumption rate
NCV	net calorific value
n	number of moles of a combustion product/mole of fuel
n-f	non-flow
p	pressure
Q, q	heat transfer (note: not a property)
R_0	universal gas constant
RAM	relative atomic mass (formerly atomic weight)
RMM	relative molecular mass (formerly molecular weight)
r	ratio
S, s	entropy
(s)	solid
s-f	steady-flow
T	absolute thermodynamic temperature (K)
t	empirical temperature ($^\circ$C); time
U, u	internal energy
V, v	volume
W, w	work transfer (note: not a property)
X	any oxygen-consuming component of a fuel
x	number of atoms of carbon/molecule fuel
y	number of atoms of hydrogen/molecule fuel
z	number of atoms of oxygen/molecule fuel
α	cut-off ratio in diesel cycle
γ	ratio of specific heat capacities
Δ	finite change

η	efficiency
ϕ	equivalence ratio

Superscripts

o	standard state of 25 ^{o}C (298.15 K) and 1 atm
'	concentration basis; theoretical; molar basis
*	algebraic sum of sensible and standard formation

Subscripts

A	air
a	atomisation
ad	adiabatic
C	change
c	combustion; corrected
cc	combustion chamber
F	fuel
f	formation
fg	saturated vapour - saturated liquid
I	initial
i	any arbitrary reactant component
j	any arbitrary product component
max	maximum
max useful	maximum useful
min	minimum
0	absolute zero temperature; environment
o	observed
P	products
p	pressure (constant)
R	reactants; reverse
r	reaction

s	stoichiometric
T	temperature (either constant, or equal initial and final); total
t	total head
V; v	volume (constant)
V	volumetric
WG	water gas

I INTRODUCTION

Combustion of mixtures of fuel with air is widely used for the conversion of chemical energy to provide heat transfer in furnaces, or
work transfer in engines. Once the components of a fuel are known
quantitatively, straightforward calculation is possible of the
stoichiometric mixture proportions (chemically balanced for complete
reaction) with air or some other oxidant. Furthermore, if the resulting products are cooled, the proportions of stable molecular
quantities are obtainable directly. Any known non-stoichiometric
reactant mixture can be handled in a similar way. In a reverse
manner, analytical data on cooled dry products of hydrocarbon combustion permit calculation of the carbon/hydrogen mass ratio of the
parent fuel, and the fuel/air mass ratio of the parent mixture.

Products that are still hot, on the other hand, are unstable,
with a tendency to dissociate back towards the reactant form. In
such cases, information on combustion kinetics is also required to
establish the degree of dissociation, and the quantities of the more
complex product mixtures of molecular, atomic and radical species
are then obtainable by iteration.

The quantity of energy released as a result of combustion can be
determined as the difference between the energy stock of the final
products and that of the initial reactants, and can be expressed as
a heat transfer in terms of a calorific value of the fuel when burnt
with the oxidant under prescribed conditions. The maximum possible
energy output from the products in the form of useful work can also
be calculated. A combination of the above information permits the
derivation of the temperature reached during combustion, and of the
efficiency of the energy-conversion process. Consequently many aspects of the combustion performance of a fuel/oxidant mixture can
be predicted from knowledge of the nature, proportions and energy
content of the mixture components. A far more efficient process of
energy conversion is available by the indirect electrochemical reaction of fuel and oxidant in a fuel cell, involving no high-temperature heat transfer at all, and this represents the ultimate in
efficiency of conversion from chemical energy to work.

TABLE 1   *RELATIVE ATOMIC AND MOLECULAR MASSES* $(C^{12} = 12)$

Element	Symbol	RAM	Approx. RAM	Compound	Symbol	RMM	Approx. RMM
Argon	Ar	39.948	40	Carbon dioxide	CO_2	44.00995	44
Carbon	C	12.01115	12	Carbon monoxide	CO	28.01055	28
Hydrogen	H	1.00797	1	Hydrogen	H_2	2.01594	2
Nitrogen	N	14.0067	14	Nitrogen	N_2	28.0134	28
Oxygen	O	15.9994	16	Oxygen	O_2	31.9988	32
Sulphur	S	32.064	32	Sulphur dioxide	SO_2	64.0628	64
				Water	H_2O	18.01534	18

2 COMBUSTION REACTANTS

The chemical elements and their oxides of interest in this study are listed in table 1 together with their relative atomic and molecular masses (RAM and RMM); for simplicity, approximate values have been adopted throughout this book. Since atmospheric air is the most common oxidant, it is also used here, and is represented in table 2. It is seen that (m) moles of oxygen are contained in (4.76m) moles of air, representing (32m) grams of oxygen and (4.31 × 32m), or (137.9m), grams of air. In some advanced applications to rocket propulsion, oxygen alone, or such alternatives as hydrogen peroxide, nitrogen oxides, nitric acid or fluorine, may be required as the oxidant, but the following treatment can be adapted for any oxidant provided its components, and their chemical behaviour, are known.

TABLE 2 COMPOSITION OF ATMOSPHERIC AIR

Component	Molar Fraction	RMM	Mass Fraction
N_2	0.781	28.0134	0.756
Ar + CO_2	0.009	(40)	0.012
'Atmospheric' N_2	0.790	(28.150)	0.768
O_2	0.210	31.9988	0.232

where ( ) signifies approximate. Thus

$$\text{'atmospheric' } N_2/O_2 \text{ ratio} = 3.76 \text{ by volume (molar)}$$

$$= 3.31 \text{ by mass}$$

$$\text{RMM air} = (28.150 \times 0.790) + (31.9988 \times 0.210)$$

$$= 28.96 \text{ approximately, allowing for traces of Ar and } CO_2$$

$$\text{Density of air} = \frac{RMM}{V_M}$$

$$= \frac{28.96}{22.4136} = 1.292 \text{ kg/m}^3 \text{ at 1 atm and 0 °C}$$

In their simplest form, fuels are represented by the great variety of either natural or manufactured hydrocarbons, which are identified individually by formulae of the $C_x H_y$ type, where x and y are integers,

3

and x is known as the carbon number. Collectively, hydrocarbons are
recognised as being one of a number of 'series', each member of a
particular series differing slightly from its adjacent fellow members
but having a general formula and structural characteristics in common.
The main series range from the straight-chain saturated *paraffins*
(alkanes) with y = 2x + 2, through the cyclic *naphthenes* (cyclanes or
cycloparaffins) and unsaturated *olefins* (alkenes) both with y = 2x,
to the highly unsaturated *acetylenes* (alkynes) with y = 2x - 2, and
eventually the cyclic *aromatics*, which, in their single 'nuclear'
configuration, have y = 2x - 6. The hydrogen content therefore falls
progressively from paraffins to aromatics, and even further when benzene rings combine to form the *polynuclear aromatics*, as with the
double ring naphthalene $C_{10}H_8$, with y = 2x - 12, and the triple ring
anthracene $C_{14}H_{10}$, with y = 2x - 18.

For an individual hydrocarbon, therefore

$$\boxed{RMM = (12x + y) \text{ approximately}} \tag{1}$$

and $\boxed{\text{carbon/hydrogen mass ratio} = (12x/y) \text{ approximately}}$ (2)

A plot of C/H mass ratio in figure 1 shows that, with the exception
of the polynuclear aromatics, values tend to 6 as x increases, since
y then tends to 2x.[1]

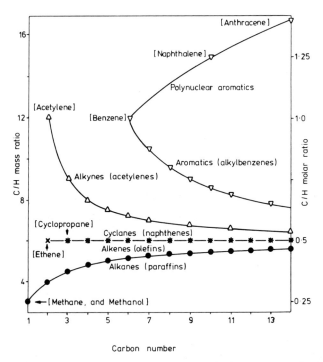

Figure 1   Carbon/hydrogen ratio of light hydrocarbons (ref. 1)

4

In practice individual hydrocarbons are rarely used as such, naturally occurring or commercially manufactured fuels generally being made up of hydrocarbon or other materials blended or compounded together in gaseous, liquid or solid form. Nevertheless, provided the main components can be identified and measured, similar calculation procedures apply.

Of the gaseous commercial fuels, natural gas is exceptional in that, in its dry state free of volatile liquids, a single hydrocarbon (methane, CH_4) is virtually the only combustible component. Petroleum gases, on the other hand, are the various blends of propane, C_3H_8, and butane, C_4H_{10}, together with minor concentrations of heavier hydrocarbons, all derived from crude petroleum. Manufactured gases used in industry are the gaseous products of carbonisation or of partial oxidation with or without water, and include coke oven gas (COG), blue water gas (BWG), carburetted water gas (CWG), coke producer gas (CPG), and blast furnace gas (BFG). Data on their relative density and volumetric composition are given in figure 2.

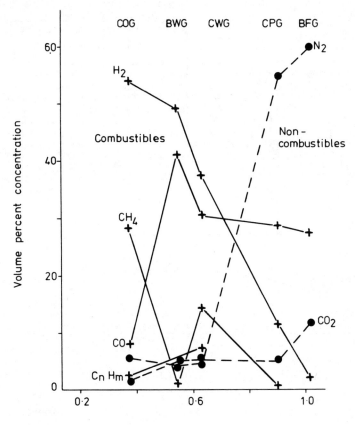

Figure 2   Composition of gaseous industrial fuels

The liquid commercial fuels are blends of petroleum-based hydro-carbons ranging from gasoline to fuel oil, which can be character-ised by such properties as relative density and boiling range, as shown in figure 3. The highly viscous coal-tar fuels are also in-cluded in the liquid fuel group, and are classified according to the temperature (°F) at which the viscosity reduces to 100 Redwood I seconds (24.1 cSt), ranging typically from CTF 50 to CTF 250.

Solid fuels, ranging from wood to anthracite, are usually classi-fied by ultimate analysis, that is, by gravimetric proportions of the major elements, as shown plotted against relative density in figure 4. The C/H mass ratio of a fuel can be determined by burning the fuel completely in a stream of oxygen, and trapping the result-ing carbon dioxide and water in weighed tubes of absorbent reagent. Figure 5 shows the resulting variation in C/H mass ratio with com-mercial fuel type.

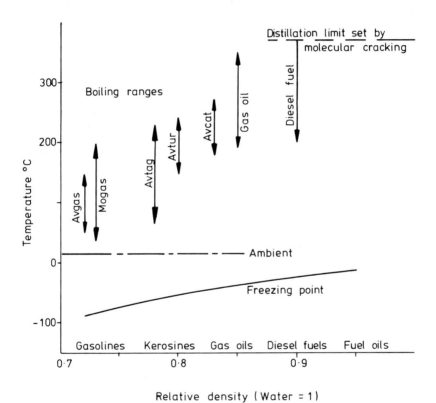

Figure 3   Boiling ranges and freezing points of liquid commercial fuels

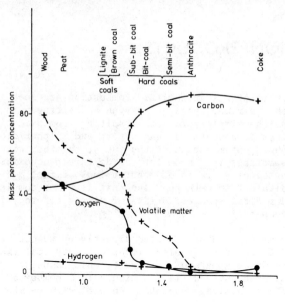

Figure 4   Composition of solid commercial fuels

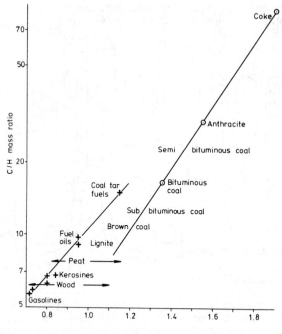

Figure 5   Carbon/hydrogen mass ratio of liquid and solid commercial fuels

3   COMBUSTION PRODUCTS

Since the most common reactants in fuel/air combustion are seen to
consist of the elements carbon, hydrogen, oxygen and nitrogen in
various combinations, the simpler products resulting from oxidation
are the complete oxides carbon dioxide and water (and sulphur dioxide
in the event of sulphur contamination of the fuel), together with
inert nitrogen and sometimes the incomplete oxide carbon monoxide
and/or unburnt hydrogen, plus traces of unburnt fuel or fuel molecular components equivalent to methane. The next level of complexity,
involving atomic and radical species, oxides of nitrogen, etc., is
usually handled by computation.

Routine product analysis can be effected directly by selective
chemical absorption of the stable cooled products. A typical example
is represented by the Orsat apparatus, figure 6, which consists of a
calibrated gas container connected via a manifold to a number of
glass U-tubes containing chemical reagents. An explosion chamber is

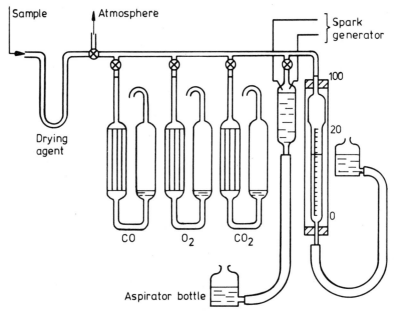

Figure 6   Orsat gas analysis apparatus

also provided for the combustion and measurement of hydrogen and fuel
molecular components. The gas container is water-jacketed to main-
tain uniform temperature, and a flexible tube is led from an unstop-
pered water-levelling aspirator bottle. The U-tubes are stacked with
glass rods to increase the wetted area for absorption. The reagents
shown are used in the following order

caustic alkali solution, for carbon dioxide
alkaline pyrogallol solution, for oxygen
ammoniacal cuprous chloride solution, for carbon monoxide
bromine water solution, for unsaturated hydrocarbons

In each case, the volume of gas absorbed is determined by measurement of the changes in sample volume using the calibrated gas container and the aspirator bottle. Finally, a known portion of the gas sample is made up to the original volume by the addition of air, and ignited in the explosion chamber. The volume of residual gas is measured, and the resulting quantity of carbon dioxide absorbed as before to determine the concentration of both methane and hydrogen. The volume of nitrogen in the sample (including errors) is determined by difference.

More accurate volumetric results are obtainable from the Haldane, Gooderham, and Bone and Wheeler types of instrument based on pressure measurements at constant volume using mercury as the displacing fluid. Alternatively, solid reagents may be used, as with soda asbestos for carbon dioxide, and magnesium perchlorate for water, contained in tubes that are weighed before and after absorption. In chromatographic methods, a mixture sample is introduced into a carrier fluid, and the components of the mixture are separated physically owing to different rates of adsorption and desorption with a stationary (solid or liquid) bed material. As each component leaves the bed in turn, the relative proportions are determined by some detecting instrument such as a katharometer or a flame ionisation detector. The former depends on the change in thermal conductivity of the carrier fluid owing to the presence of the desorbed component, whereas the latter senses the substantial increase in ion concentration in a hydrogen/oxygen flame owing to the presence of carbon compounds. The components can be identified by comparing them with known samples.

Many other techniques are available for the determination of fuel or product components, or of mixture strength in the case of fuels of a known type, based on the accurate assessment of such properties as infrared absorptivity, paramagnetism, sonic velocity, chemiluminescence and relative density. However, examination questions dealing with gas analysis as a general aspect of combustion are usually concerned with volumetric analysis of the Orsat type rather than the more sophisticated techniques available.

4 PROPORTIONS OF REACTANTS AND COOLED PRODUCTS

The stoichiometric mixture proportions of any fuel with air may be expressed in terms of the air/fuel ratio $(A/F)_s$, or its inverse fuel/air ratio $(F/A)_s$, on either a gravimetric or volumetric (molar) basis. The proportions of a non-stoichiometric mixture are often expressed in terms of percentage of fuel weakness or richness, that is, of excess or insufficiency of air. Alternative expressions are the actual air/fuel ratio A/F, fuel/air ratio F/A, or the equivalence ratio given by

$$\boxed{\text{equivalence ratio} = \phi = \frac{F/A}{(F/A)_s}} \tag{3}$$

$$= \frac{(A/F)_s}{A/F} = \frac{m_s}{m} = \frac{M_s}{M}$$

where m_s = stoichiometric moles of oxygen per mole of fuel, m = actual moles of oxygen per mole of fuel, M_s = stoichiometric moles of oxygen per gram of fuel and M = actual moles of oxygen per gram of fuel. The above definitions of M are used since the SI mole is the quantity of material whose mass is equal to the RMM of the material in grams (not kilograms). In the fuel-rich case, it follows that $m < m_s$, and $\phi > 1$. The converse is true in the fuel-weak case.

The concentrations of combustion products may be expressed on a wet basis as volume percentages of the total products. For concentrations on a dry basis, the quantity of water is excluded from the total products. The basic oxidation equations for carbon and hydrogen are

$$C + 0.5O_2 \longrightarrow CO$$
$$CO + 0.5O_2 \longrightarrow CO_2$$
$$\left.\right\} \quad C + O_2 \longrightarrow CO_2$$
$$H_2 + 0.5O_2 \longrightarrow H_2O$$

The above molar quantities can be converted to a gravimetric basis as follows

$$\boxed{\text{mass material} = \text{moles material} \times \text{RAM or RMM material}} \tag{4}$$

Hence (1×12)g C burn with (1×32)g O_2 to produce (1×44)g CO_2, and (1×2)g H_2 burn with (0.5×32)g O_2 to produce (1×18)g H_2O.

At first sight, carbon monoxide and possibly hydrogen would be expected to be included in the products from a fuel-rich mixture,

10

with excess oxygen when fuel-weak. However, at high temperatures (above about 1800 K), products become so energetic that they dissociate by reverse reactions to give CO, H_2, H, O, OH, etc., as discussed in chapter 5. Although dissociation will have ceased during the subsequent cooling of the products, the oxidation of some of the unburnt species may be inhibited owing to chilling, in which case both rich- and weak-type products will co-exist, and appear in the analysis. For this reason, the following *general combustion equation,* on a molar basis, is envisaged to allow for all major molecular components and for any mixture strength

$$1(fuel) + m(O_2 + 3.76N_2)$$
$$= n_1CO_2 + n_2H_2O + n_3CO + n_4H_2 + n_5O_2 + n_6N_2 \qquad (5)$$

Sulphur, free carbon and other materials and their oxides may be included as appropriate. Where approximate results are acceptable, term n_5O_2 is frequently disregarded in the fuel-rich case, and term n_3CO in the fuel-weak case. Furthermore, since hydrogen is more reactive than carbon, it is common practice to disregard the minor term n_4H_2, and to assume that, in the fuel-rich case, the available oxygen combines preferentially to form water, the remainder being shared between carbon dioxide and carbon monoxide. Information on the *relative* quantities of these products permits solution.

The 'molar balance (oxygen)' method of calculation is based either on (m) moles of oxygen supplied per unit mole of fuel, or on (M) moles of oxygen per unit mass of fuel depending, respectively, on whether the fuel component analysis is given on a volumetric or gravimetric basis. This method lends itself more widely to non-stoichiometric mixtures than the tabular method using a comprehensive reaction table and, being based on oxygen rather than air, is also more readily adaptable to other oxidants.

4.1   INDIVIDUAL HYDROCARBONS

For a hydrocarbon fuel C_xH_y, in the absence of chilled products of dissociation, the *stoichiometric combustion equation* on a molar basis is

$$C_xH_y + m_s(O_2 + 3.76N_2) = n_1CO_2 + n_2H_2O + n_6N_2$$

[Individual hydrocarbon-derived molecules containing oxygen $C_xH_yO_z$ (for example, alcohols) can also appear here.] Although the values of m_s and n are immediately apparent from inspection, the molar balance method of solution is followed systematically here as practice for the more complex cases later.

Molar balance gives

carbon balance: $n_1 = x$

11

hydrogen balance: $n_2 = y/2$

oxygen balance: $n_1 + 0.5n_2 = \boxed{m_s = (x + y/4)}$ (6)

nitrogen balance: $n_6 = 3.76m_s$

Hence

volumetric $(A/F)_s = 4.76m_s$

thus $\boxed{\text{volumetric } (A/F)_s = 4.76(x + y/4)}$ (7)

and gravimetric $(A/F)_s = \dfrac{(28.96)}{(12x + y)} \, 4.76(x + y/4)$ approximately

thus $\boxed{\text{gravimetric } (A/F)_s = \dfrac{137.9(x + y/4)}{12x + y} \text{ approximately}}$ (8)

Combustion product proportions are then found by construction of of a products table showing summations of the product moles, and evaluation of expressions of the type

$$\boxed{\text{volume \% product } j = 100 \, \frac{\text{mol } j}{\Sigma \text{ mol products}}}$$ (9)

as follows.

Product	mol	volume % wet	volume % dry
CO_2	n_1		
N_2	n_6		
Σ dry =		–	–
H_2O	n_2		–
Σ wet =		100	100

The totals in the two final columns are usually considered sufficiently accurate at $100 \pm 0.2\%$.

It follows that

$$\text{mass \textit{wet} products/unit mass fuel} = \frac{A + F}{F}$$

$$= A/F + 1$$

where A and F are both masses, and that

12

mass *dry* products
/unit mass fuel = A/F + 1 - $\dfrac{18n_2}{\text{RMM fuel (mol fuel)}}$

Also mol *wet* products/mol fuel = $\dfrac{\Sigma n}{\text{mol fuel}}$

and mol *dry* products/mol fuel = $\dfrac{\Sigma n - n_2}{\text{mol fuel}}$

[see example 1]

For any non-stoichiometric mixture of a hydrocarbon in air, the general combustion equation on a molar basis, allowing for chilled products of dissociation, is

$$C_x H_y + m(O_2 + 3.76N_2)$$

$$= n_1 CO_2 + n_2 H_2 O + n_3 CO + n_4 H_2 + n_5 O_2 + n_6 N_2$$

Molar balance gives

carbon balance: $n_1 + n_3 = x$

hydrogen balance: $n_2 + n_4 = y/2$

oxygen balance: $n_1 + 0.5n_2 + 0.5n_3 + n_5 = m$

nitrogen balance: $n_6 = 3.76m$

[see example 2]

If the term $n_4 H_2$ together with terms $n_5 O_2$ or $n_3 CO$ are not to be deleted, as indicated on p.11, two further equations are required for solution, and these may be provided by ratios of n values from partial analysis of the products. Complete combustion product calculations then follow, using a products table as in the non-chilled stoichiometric case.

Figure 7a shows variations in carbon dioxide, carbon monoxide and unburnt hydrocarbons expressed on a dry volumetric basis for a typical gasoline burning in air within a spark-ignition piston engine. Owing to mixing imperfections, local chilling, etc., small quantities of carbon monoxide and unburnt hydrocarbons persist into the fuel-weak region. For furnaces and compression-ignition piston engines that operate on the fuel-weak side of stoichiometric, product concentrations may be more conveniently plotted as an Ostwald diagram, figure 7b, with percentage carbon dioxide against percentage oxygen, including minor concentrations of carbon monoxide and showing the percentage air used, or alternatively as a nomogram of some kind incorporating data on density, humidity, loss due to carbon monoxide, boiler efficiency, etc.

13

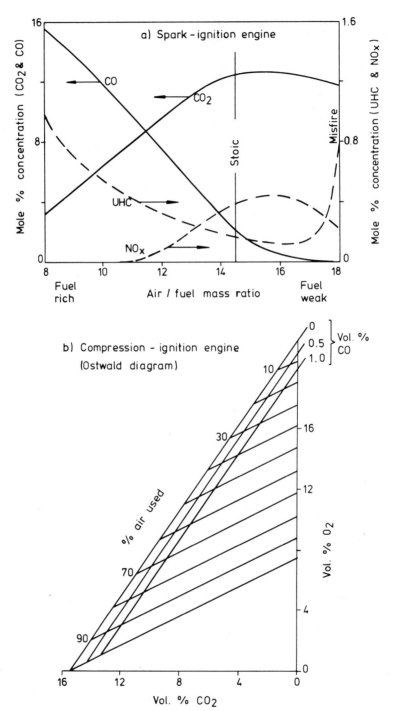

Figure 7   Concentrations of products of hydrocarbon combustion in piston engines

14

4.2 VOLUMETRIC FUEL BLENDS

Gaseous, and most liquid, fuel blends are described in terms of volumetric (that is, molar) fractions of molecular components. For one mole of a representative fuel blend consisting of, for example

$$(aC_xH_y + bH_2 + cCO + dO_2 + eN_2)$$

where a, b, c, etc., are molar fractions, the over-all value of m_s for the blend is found by multiplying the individual value of m_s for each oxygen-consuming component by its molar fraction, and summing, as follows.

Component: C_xH_y H_2 CO O_2

$$\boxed{m_s \text{ for fuel blend} = (x + y/4)a + 0.5b + 0.5c - d} \qquad (10)$$

As before

$$\text{volumetric } (A/F)_s = 4.76m_s$$

and $$\text{gravimetric } (A/F)_s = \frac{137.9m_s}{RMM \text{ fuel}}$$

thus

$$\boxed{\begin{aligned} &\text{gravimetric } (A/F)_s \qquad\qquad\qquad\qquad\qquad (11)\\[4pt] &= 137.9 \; \frac{\Sigma(m_s \text{ 'X'} \times \text{molar fraction 'X'}) - \text{molar fraction } O_2}{\Sigma(RMM \text{ fuel component} \times \text{molar fraction fuel component})} \end{aligned}}$$

approximately, where 'X' is an oxygen-consuming fuel component.

For any non-stoichiometric mixture, the general combustion equation on a molar basis is

$$(aC_xH_y + bH_2 + cCO + dO_2 + eN_2) + m(O_2 + 3.76N_2)$$

$$= n_1CO_2 + n_2H_2O + n_3CO + n_4H_2 + n_5O_2 + n_6N_2$$

carbon balance: $n_1 + n_3 = ax + c$

hydrogen balance: $n_2 + n_4 = 0.5ay + b$

oxygen balance: $n_1 + 0.5n_2 + 0.5n_3 + n_5 = 0.5c + d + m$

nitrogen balance: $n_6 = e + 3.76m$

As outlined on p.11, elimination of n_4 with n_5 or n_3, or knowledge of ratios of these quantities, permits solution for m and all values of n. Hence

15

$$\text{volumetric A/F} = 4.76m$$

$$\text{and gravimetric A/F} = \frac{137.9m}{\text{RMM fuel}}$$

as above. Tabulated combustion products calculations then follow, as before

[see example 3]

4.3 GRAVIMETRIC FUEL BLENDS

Solid fuels are usually described in terms of gravimetric fractions of elemental combustible components, together with any contaminant water, ash and sulphur. Unit mass of fuel is therefore taken as a basis, and the mass fractions of components converted to molar fractions per unit mass of fuel by division with the relevant value of RAM or RMM. For unit mass of a representative solid fuel consisting of

$$(aC + bH_2 + cO_2 + dN_2 + eH_2O + fS + gAsh)$$

where a, b, c, etc., are mass fractions, molar/mass conversion gives

$$\left(\frac{aC}{12} + \frac{bH_2}{2} + \frac{cO_2}{32} + \frac{dN_2}{28} + \frac{eH_2O}{18} + \frac{fS}{32} \right)$$

approximately, ignoring the inorganic ash that appears on both sides of the combustion equation. In SI, the molar unit of quantity is based on the *gram*, and the unit of mass on the *kilogram*, hence the values of the molar/mass fractions are identical in units of mol/g and kmol/kg.

The over-all value of M_s for the fuel is found, as before, by multiplying the individual value of m_s for each oxygen-consuming component by its molar/mass fraction, and summing, as follows.

Component: C H_2 S O_2

$$M_s = \frac{1a}{12} + \frac{0.5b}{2} + \frac{f}{32} - \frac{c}{32} \qquad (12)$$

Hence

$$\text{gravimetric } (A/F)_s = 4.31(32M_s) \text{ approximately}$$

$$= 137.9M_s \text{ approximately}$$

$$\text{gravimetric } (A/F)_s$$
$$= 137.9 \left[\Sigma \left(\frac{m_s \text{ 'X' } \times \text{ mass fraction 'X'}}{\text{RMM 'X'}} \right) - \frac{\text{mass fraction } O_2}{32} \right] \qquad (13)$$

16

approximately, where 'X' is an oxygen-consuming fuel component.

[see example 4]

For any non-stoichiometric mixture, the general procedure is similar to that in section 4.2, but with m replaced by M. Allowance can be made for any free carbon in the products. The final expression is then

gravimetric A/F = 137.9M

approximately. Tabulated combustion products calculations then follow, as before

[see example 5]

4.4   GAS ABSORPTION ANALYSIS

When used for gaseous combustion products, the analysis is usually restricted to carbon dioxide, carbon monoxide and oxygen only, with nitrogen obtained by difference, as in the Orsat method.[2] However, provision is made to measure hydrogen, methane and unsaturated hydrocarbons in addition and, for completeness, these are included in the following treatment.

For an initial volume of sample of 100 cm^3, let

volume after CO_2 absorption = V_1

volume after O_2 absorption = V_2

volume after CO absorption = V_3

volume after unsaturated absorption = V_4

volume after explosion of 20 cm^3 of remaining gas with 80 cm^3 of air = V_5

volume after subsequent CO_2 absorption = V_6

The explosion reactions are as follows

$$aCH_4(g) + 2aO_2(g) = aCO_2(g) + 2aH_2O(l)$$

and $bH_2(g) + 0.5bO_2(g) = bH_2O(l)$

giving reductions of 2a and 1.5b respectively on condensation. Hence

$$a = V_5 - V_6$$

and $2a + 1.5b = 100 - V_5$

so that

$$b = \frac{100 - V_5 - 2a}{1.5}$$

17

The full analysis on a dry basis then follows.

Volume % CO_2 = 100 - V_1

Volume % O_2 = V_1 - V_2

Volume % CO = V_2 - V_3

Volume % C_xH_{2x} = V_3 - V_4

Volume % CH_4 = $\dfrac{a}{20} V_4$

Volume % H_2 = $\dfrac{b}{20} V_4$

Volume % N_2 (+ errors) = 100 - sum of above

[see example 6]

4.5   FUEL AND MIXTURE CALCULATION FROM PRODUCT ANALYSIS

The following treatment represents the above procedures used in a reverse sense to determine the carbon/hydrogen mass ratio of a hydrocarbon fuel, and the air/fuel mass ratio of the mixture, from product analysis data derived on a cooled dry basis.

4.5.1   Carbon/Hydrogen Mass Ratio of Fuel

From the general combustion equation on a molar basis for a hydrocarbon

$$C_xH_y + m(O_2 + 3.76N_2)$$
$$= n_1CO_2 + n_2H_2O + n_3CO + n_4H_2 + n_5O_2 + n_6N_2$$

the cooled dry volumetric products analysis provides values of the type

$$\text{volume \% } CO_2 = \frac{100n_1}{\Sigma \text{ mol dry products}}$$

and so on, where Σ mol dry products excludes n_2. The absolute values of n and Σ mol dry products are unknown, but the data give the relationships between the values of n, hence each can be expressed in terms of n_1, as follows.

carbon balance: $x = n_1 + n_3 = n_1 + \left(\dfrac{n_3}{n_1}\right)n_1$

$$= \text{function 1 of } n_1 = f_1(n_1)$$

hydrogen balance: $y = 2n_2 + 2n_4 = 2n_2 + 2\left(\dfrac{n_4}{n_1}\right)n_1 = 2n_2 + f_2(n_1)$

oxygen balance: $m = n_1 + 0.5n_2 + 0.5n_3 + n_5$

18

$$= \left[1 + 0.5\left(\frac{n_3}{n_1}\right) + \left(\frac{n_5}{n_1}\right)\right]n_1 + 0.5n_2$$

$$= f_3(n_1) + 0.5n_2$$

nitrogen balance: $n_6 = 3.76m = \left(\frac{n_6}{n_1}\right)n_1$

Thus $m = \frac{1}{3.76}\left(\frac{n_6}{n_1}\right)n_1 = f_4(n_1)$

From the last two expressions above for m

$$n_2 = 2[f_4(n_1) - f_3(n_1)] = f_5(n_1)$$

Hence

$$C/H \text{ mass ratio} = \frac{12x}{y}$$

$$= \frac{12f_1(n_1)}{2f_5(n_1) + f_2(n_1)}$$

which resolves by cancellation of n_1.

4.5.2 Air/Fuel Mass Ratio of Mixture

$$\text{Actual gravimetric A/F} = \frac{137.9m}{12x + y}$$

$$= \frac{137.9f_4(n_1)}{12f_1(n_1) + 2f_5(n_1) + f_2(n_1)}$$

from which, again, n_1 cancels to give a solution.

[see example 7]

4.6 PHYSICAL CHARACTERISTICS OF MIXTURES AND PRODUCTS

On the basis of one mole of fuel with air

$$\Sigma \text{ mol gaseous reactants} = (1 + 4.76m) \text{ mol}$$

if the fuel temperature is greater than the fuel boiling point, (or the fuel is considered to be theoretically in the gaseous phase at standard conditions irrespective of its boiling point), whereas

$$\Sigma \text{ mol gaseous reactants} = 4.76m \text{ mol}$$

if the fuel temperature is less than the fuel boiling point. From the equation of state for ideal gases

$$pV = nR_0T$$

where n = number of moles of gas, and R_0 = universal gas constant = 8.3143 J/mol K. Hence

$$\text{volume of reactants} = \frac{8.3143T_R}{p_R} \; (\Sigma \text{ mol gaseous reactants})$$

$$\text{m}^3/\text{mol fuel}$$

$$\text{and} \quad \text{volume of products} = \frac{8.3143T_P}{p_P} \; (\Sigma \text{ mol gaseous products})$$

$$\text{m}^3/\text{mol fuel}$$

hence the volume change resulting from combustion is given by

$$\Delta V_C = 8.3143 \left[\frac{T_P}{p_P} \; (\Sigma \text{ mol gaseous products}) - \frac{T_R}{p_R} \; (\Sigma \text{ mol gaseous reactants}) \right] \text{m}^3/\text{mol fuel} \tag{14}$$

Although, for simplicity, dry products analysis is based on the complete removal of water, some water vapour exists in the product gases 'in the form of humidity. Hence chemical drying agents would be required after cooling to ambient if the last traces of moisture are to be removed. The humidity of the products is derived using the Saturated Water and Steam Tables [3] in the following manner.

If the temperature of the products is greater than the saturation temperature corresponding to the *total* pressure of the products, the temperature of the water product is clearly much greater than the saturation temperature corresponding to the *partial* pressure of the water, since saturation temperature decreases with pressure. In such a case, the water exists completely as vapour, and (Σ mol gaseous products) includes n_2.

The quantity of water condensing at some lower product temperature, for example ambient, is also determined by use of the saturated water and steam tables, as follows.

$$\frac{V_{steam}}{V} = \frac{n_{steam}}{(n_{steam} + \Sigma \text{ mol dry gaseous products})} = \frac{p_{steam}}{p}$$

where p_{steam} = partial pressure of saturated steam at product temperature. Thus

$$n_{steam} = \frac{p_{steam}}{(p - p_{steam})} \; (\Sigma \text{ mol dry gaseous products})$$

Hence

$$H_2O \text{ condensed} = (n_2 - n_{steam}) \text{ mol/mol fuel}$$

20

$$= \frac{18(n_2 - n_{steam})}{RMM\ fuel}\ mass/mass\ fuel$$

At dew point of products

$$p_{steam} = p\left(\frac{n_2}{n_2 + \Sigma\ mol\ dry\ gaseous\ products}\right)$$

$$= p\left(\frac{n_2}{\Sigma\ mol\ wet\ gaseous\ products}\right)$$

From saturated water and steam tables

dew point = saturation temperature corresponding to p_{steam}

[see example 8]

4.7 EXAMPLES

The general approach is seen to consist of writing the appropriate combustion equation in molar terms, and solving by means of molar balances of carbon, hydrogen, oxygen, nitrogen, etc., using any partical analytical data provided. A products table is then constructed to determine systematically the volumetric percentages. As indicated in section 4.1, the values of n and m_s are often immediately apparent in the simpler examples, but the molar balance method of solution is followed throughout to give practice in its use.

Example 1

For the gaseous fuel methane, CH_4, determine the stoichiometric air/ fuel ratios by volume and by mass, together with the wet and dry products of combustion, and also the mass of dry products per unit mass of fuel.

From section 4.1, the molar stiochiometric combustion equation is

$$CH_4 + m_s(O_2 + 3.76N_2) = n_1CO_2 + n_2H_2O + n_6N_2$$

carbon balance: $n_1 = x = 1$

hydrogen balance: $n_2 = y/2 = 2$

oxygen balance: $n_1 + n_2/2 = m_s = (x + y/4) = 2$

nitrogen balance: $n_6 = 3.76m_s = 7.52$

Hence

volumetric $(A/F)_s = 4.76m_s = 9.52$

and gravimetric $(A/F)_s = \frac{137.9(2)}{(12 + 4)} = 17.24$ approximately

21

Product	mol	Vol % wet	Vol % dry
CO_2	1	9.51	11.74
N_2	7.52	71.48	88.26
Σ dry =	8.52	-	-
H_2O	2	19.01	-
Σ wet =	10.52	100.00	100.00

$$\text{Mass of dry products per unit mass of fuel} = (A/F)_s + 1 - \frac{18n_2}{\text{RMM fuel (mol fuel)}}$$

$$= 17.24 + 1 - \frac{18 \times 2}{16 \times 1}$$

$$= 15.99 \text{ kg/kg fuel}$$

Example 2

Determine the approximate air/fuel ratios by mass, and the wet and dry products of combustion, for weak and rich mixtures of ethanol (C_2H_5OH), which have respectively (a) 10% excess air, assuming no carbon monoxide or hydrogen in products, and (b) 10% insufficient air, assuming no oxygen or hydrogen in products.

From section 4.1

$$m_s = (x + y/4 - z/2) = 2 + 1.5 - 0.5 = 3$$

RMM fuel = 24 + 6 + 16 = 46

Thus gravimetric $(A/F)_s = \dfrac{137.9(3)}{46} = 8.99$ approximately

(a) *10% excess air* (Note: $\phi = m_s/m = 1/1.1 = 0.91$)

Gravimetric $(A/F) = 1.1(A/F)_s = 1.1(8.99) = 9.89$ approximately

and $m = 1.1m_s = 1.1(3) = 3.3$

The general combustion equation on a molar basis is

$$C_2H_5OH + m(O_2 + 3.76N_2) = n_1CO_2 + n_2H_2O + n_5O_2 + n_6N_2$$

carbon balance: $n_1 = x = 2$ (as in stoichiometric case)

hydrogen balance: $n_2 = y/2 = 3$ (as in stoichiometric case)

oxygen balance: $n_5 = 0.1m_s = 0.3$

nitrogen balance: $n_6 = 3.76m = 3.76(3.3) = 12.41$

Product	mol	Vol % wet	Vol % dry
CO_2	2	11.29	13.60
O_2	0.3	1.69	2.04
N_2	12.41	70.07	84.36
Σ dry =	14.71	-	-
H_2O	3	16.94	-
Σ wet =	17.71	99.99	100.00

(b) *10% insufficient air* (Note: $\phi = m_s/m = 1/0.9 = 1.11$)

Gravimetric $(A/F) = 0.9(A/F)_s = 0.9(8.99) = 8.09$

and $m = 0.9m_s = 0.9(3) = 2.7$

The general combustion equation, on a molar basis is

$C_2H_5OH + m(O_2 + 3.76N_2) = n_1CO_2 + n_2H_2O + n_3CO + n_6N_2$

carbon balance: $n_1 + n_3 = x = 2$

hydrogen balance: $n_2 = y/2 = 3$ (as in stoichiometric case)

oxygen balance: $z/2 + m = n_1 + 0.5n_2 + 0.5n_3$

$$= (n_1 + n_3) - 0.5n_3 + 0.5n_2$$

thus $n_3 = 2[(n_1 + n_3) + 0.5n_2 - m - z/2]$

$$= 2(2 + 1.5 - 2.7 - 0.5)$$

$$= 0.6$$

and $n_1 = 2 - n_3 = 1.4$

nitrogen balance: $n_6 = 3.76m = 3.76(2.7) = 10.15$

Product	mol	Vol % wet	Vol % dry
CO_2	1.4	9.24	11.52
CO	0.6	3.96	4.94
N_2	10.15	67.00	83.54
Σ dry =	12.15	-	-
H_2O	3	19.80	-
Σ wet =	15.15	100.00	100.00

23

Example 3

A sample of coal-gas has the following volumetric percentage composition

methane 20, butene 2, hydrogen 50, carbon monoxide 18, oxygen 1, nitrogen 5, carbon dioxide 4

Determine the stoichiometric air/fuel ratio by volume, and the wet and dry products of combustion at 20% excess air, assuming no carbon monoxide or hydrogen in the products.

From section 4.2

$$m_s = \Sigma (m_s \text{ 'X'} \times \text{molar fraction 'X'}) - \text{molar fraction } O_2$$

where 'X' is an oxygen-consuming fuel component.

Component: CH_4 C_4H_8 H_2 CO O_2

$$m_s = 2(0.2) + 6(0.02) + 0.5(0.5) + 0.5(0.18) - 0.01$$

$$= 0.4 + 0.12 + 0.25 + 0.09 - 0.01$$

$$= 0.85 \text{ mol/mol fuel}$$

Thus volumetric $(A/F)_s = 4.76 m_s = 4.76(0.85) = 4.05$

20% excess air

$$m = 1.2 m_s = 1.2(0.85) = 1.02 \text{ mol/mol fuel}$$

With no carbon monoxide or hydrogen as products, the combustion equation on a molar basis is as follows

$$(0.2CH_4 + 0.02C_4H_8 + 0.5H_2 + 0.18CO + 0.01O_2 + 0.05N_2 +$$

$$0.04CO_2) + 1.02(O_2 + 3.76N_2) = n_1CO_2 + n_2H_2O + n_5O_2 + n_6N_2$$

carbon balance: $n_1 = 0.2 + 0.08 + 0.18 + 0.04 = 0.5$

hydrogen balance: $n_2 = 0.4 + 0.08 + 0.5 = 0.98$

oxygen balance: $n_1 + 0.5n_2 + n_5 = 0.09 + 0.01 + 0.04 + 1.02$
$$= 1.16$$

thus $n_5 = 1.16 - 0.5 - 0.49 = 0.17$

nitrogen balance: $n_6 = 0.05 + 3.76(1.02) = 3.89$

Product	mol	Vol % wet	Vol % dry
CO_2	0.5	9.03	10.96
O_2	0.17	3.07	3.73
N_2	3.89	70.22	85.31
Σ dry =	4.56	-	-
H_2O	0.98	17.69	-
Σ wet =	5.54	100.01	100.00

Example 4

A sample of anthracite has the following gravimetric percentage composition

carbon 90, hydrogen 3, oxygen 2, nitrogen 1, sulphur 1, ash 3

Determine the stoichiometric air/fuel mass ratio, together with the air/fuel mass ratio and products of combustion at 20% excess air, assuming no carbon monoxide or hydrogen in the products.

From section 4.3

$$M_s = \Sigma \left(\frac{m_s \ 'X' \times \text{mass fraction } 'X'}{\text{RMM } 'X'} \right) - \frac{\text{mass fraction } O_2}{32}$$

where 'X' is an oxygen-consuming fuel component. Thus

$$M_s = \frac{1(0.9)}{12} + \frac{0.5(0.03)}{2} + \frac{0.01}{32} - \frac{0.02}{32}$$

$$= 1(0.075) + 0.5(0.015) + 0.0003 - 0.0006$$

$$= 0.0822 \text{ mol/g fuel (or kmol/kg fuel)}$$

and gravimetric $(A/F)_s = 137.9(0.0822)$

$$= 11.34$$

20% excess air

Gravimetric A/F = 1.2(11.34) = 13.61

The combustion equation, on a basis of mol/g (or kmol/kg) of fuel, is

$$(0.075C + 0.015H_2 + 0.0006O_2 + \frac{0.01}{28} N_2 + 0.0003S) + 1.2M_s$$

$$(O_2 + 3.76N_2) = n_1CO_2 + n_2H_2O + n_5O_2 + n_6N_2 + n_7SO_2$$

25

carbon balance: $n_1 = 0.075$

hydrogen balance: $n_2 = 0.015$

sulphur balance: $n_7 = 0.0003$

oxygen balance: $n_5 = 0.2M_s = 0.0164$

nitrogen balance: $n_6 = (1.2 \times 0.0822 \times 3.76) + \dfrac{0.01}{28}$

$$= 0.3708 + 0.0004$$

$$= 0.3712$$

Product	mol	Vol % wet	Vol % dry
CO_2	0.075	15.69	16.20
O_2	0.0164	3.43	3.54
N_2	0.3712	77.67	80.19
SO_2	0.0003	0.06	0.06
Σ dry =	0.4629	-	-
H_2O	0.015	3.14	-
Σ wet =	0.4779	99.99	99.99

Example 5

A sample of hard coal has the following gravimetric percentage composition

    carbon 88, hydrogen 5, oxygen 4, ash 3

The results from a combustion test show a partial dry products volumetric analysis of 15% carbon dioxide, 2% carbon monoxide and 1% hydrogen, with the ash containing 0.01g unburnt carbon per gram of fuel supplied. Determine the air/fuel mass ratio, the equivalence ratio, and the wet products composition.

From section 4.3

$$\text{gravimetric } (A/F)_s = 137.9 \left[\Sigma \left(\frac{m_s \text{ 'X'} \times \text{mass fraction 'X'}}{\text{RMM 'X'}} \right) - \frac{\text{mass fraction } O_2}{32} \right]$$

where 'X' is an oxygen-consuming fuel component. Thus

26

$$\text{gravimetric } (A/F)_s = 137.9\left[\frac{1(0.88)}{12} + \frac{0.5(0.05)}{2} - \frac{0.04}{32}\right]$$

$$= 137.9[1(0.0733) + 0.5(0.025) - 0.00125]$$

$$= 137.9(0.08455)$$

$$= 11.66$$

The general combustion equation, on a basis of mol/g (or kmol/kg) of fuel, is

$$(0.0733C + 0.025H_2 + 0.001250_2) + M(O_2 + 3.76N_2) = n_1CO_2 +$$

$$n_2H_2O + n_3CO + n_4H_2 + n_5O_2 + n_6N_2 + \frac{0.01C}{12}$$

carbon balance: $n_1 + n_3 = 0.0733 - 0.0008 = 0.0725$ (A)

hydrogen balance: $n_2 + n_4 = 0.025$

oxygen balance: $n_1 + 0.5n_2 + 0.5n_3 + n_5 = 0.00125 + M$

thus $n_5 = M + 0.00125 - 0.5n_2 - (n_1 + n_3) + 0.5n_3$

$$= M - 0.07125 - 0.5n_2 + 0.5n_3 \qquad\qquad (B)$$

nitrogen balance: $n_6 = 3.76M$

Product analysis gives

$$n_3/n_1 = 2/15 = 0.1333$$

and $n_4/n_1 = 1/15 = 0.0667$

From (A)

$$n_1 = \frac{0.0725}{1 + n_3/n_1} = \frac{0.0725}{1.1333} = 0.0640$$

and $n_3 = 0.0725 - 0.0640 = 0.0085$

Also $n_4 = 0.0667(0.0640) = 0.0043$

and $n_2 = 0.025 - 0.0043 = 0.0207$

Since

$$\frac{n_1}{n_1 + n_3 + n_4 + n_5 + n_6} = 0.15$$

$$n_5 = \frac{n_1}{0.15} - (n_1 + n_3) - n_4 - n_6$$

$$= \frac{0.0640}{0.15} - 0.0725 - 0.0043 - 3.76M$$

$$= 0.3499 - 3.76M \qquad\qquad\text{(C)}$$

From (B)

$$n_5 = M - 0.07125 - 0.5(0.0207) + 0.5(0.0085)$$

$$= M - 0.0774 \qquad\qquad\text{(D)}$$

Equating (C) and (D)

$$M = 0.4273/4.76$$

$$= 0.0898$$

Thus gravimetric A/F = 137.9(0.0898)

$$= 12.38$$

Hence

$$\phi = \frac{(A/F)_s}{A/F} = \frac{11.66}{12.38} = 0.942$$

that is, a fuel-weak mixture. From (D)

$$n_5 = 0.0898 - 0.0774$$

$$= 0.0124$$

and $n_6 = 3.76M = 0.3376$

Product	mol	Vol % wet
CO_2	0.0640	14.30
H_2O	0.0207	4.63
CO	0.0085	1.90
H_2	0.0043	0.96
O_2	0.0124	2.77
N_2	0.3376	75.44
Σ wet =	0.4475	100.00

Example 6

Determine the complete volumetric dry products of combustion from the following results of an Orsat-type analysis.

Initial volume of sample = 100 cm^3

Volume after CO_2 absorption = V_1 = 86.0 cm^3

28

Volume after O_2 absorption = V_2 = 83.7 cm^3

Volume after CO absorption = V_3 = 80.7 cm^3

Volume after explosion of 20 cm^3 gas with 80 cm^3 air = V_5 = 99.5 cm^3

Volume after subsequent CO_2 absorption = V_6 = 99.4 cm^3

From section 4.4, since no unsaturateds were measured, volume after unsaturated absorption = V_4 = V_3 = 80.7 cm^3.

$a = V_5 - V_6 = 0.1$

$b = \dfrac{100 - 99.5 - 0.2}{1.5} = 0.2$

Product	Vol % dry	
CO_2	$100 - V_1$ =	14.0
O_2	$V_1 - V_2$ =	2.3
CO	$V_2 - V_3$ =	3.0
C_xH_{2x}	$V_3 - V_4$ =	0
CH$_4$	$aV_4/20$ =	0.4
H_2	$bV_4/20$ =	0.8
N_2 + losses, by difference =		79.5
	Total =	100.0

Example 7

The dry products of combustion of a hydrocarbon/air mixture have the following volumetric percentage analysis

carbon dioxide 12.5, carbon monoxide 2.5, nitrogen 85.0

Determine the C/H mass ratio of the fuel, and the air/fuel mass ratio of the reactant mixture assuming that no fuel remains unreacted.

From section 4.5.1, the general combustion equation on a molar basis is

$$C_xH_y + m(O_2 + 3.76N_2) = n_1CO_2 + n_2H_2O + n_3CO + n_6N_2$$

Product analysis gives

n_3/n_1 = 2.5/12.5, therefore $n_3 = 0.2n_1$

n_6/n_1 = 85.0/12.5, therefore $n_6 = 6.8n_1$

carbon balance: $x = n_1 + n_3 = 1.2n_1$

hydrogen balance: $0.5y = n_2$, therefore $y = 2n_2$

oxygen balance: $m = n_1 + 0.5n_2 + 0.5n_3 = 1.1n_1 + 0.5n_2$

nitrogen balance: $n_6 = 3.76m$, therefore $m = n_6/3.76$

$$= 6.8n_1/3.76$$

$$= 1.81n_1$$

Equating the above two expressions for m

$$1.1n_1 + 0.5n_2 = 1.81n_1$$

thus $n_2 = \dfrac{0.71n_1}{0.5} = 1.42n_1$

Therefore

C/H mass ratio of fuel $= \dfrac{12x}{y} = \dfrac{12(1.2n_1)}{2.84n_1} = 5.07$

From section 4.5.2

actual gravimetric A/F $= \dfrac{(28.96)}{(12x + y)} \, 4.76m$

$$= \dfrac{137.9 \times 1.81n_1}{(12 \times 1.2 + 2.84)n_1}$$

$$= 14.48$$

Example 8

Determine the volumes per kilogram of fuel of a stoichiometric mixture with air of ethanol (bp = 78 °C) at temperature 80 °C and pressure 1.013 bar, and of its products after cooling to 150 °C at this pressure. Find also the mass of water condensed per kilogram of fuel when the products are cooled further to 15 °C, and find the dewpoint at this pressure.

The stoichiometric combustion equation on a molar basis (as in example 2, or by inspection) is

$$C_2H_5OH + 3(O_2 + 3.76N_2) = 2CO_2 + 3H_2O + 11.28N_2$$

From section 4.6, since reactant temperature > fuel bp

Σ mol gaseous reactants $= 1 + 4.76(3) = 15.28$ mol/mol fuel

Thus volume of reactants $= \dfrac{8.3143T}{p} \left(\dfrac{\Sigma \text{ mol gaseous reactants}}{\text{RMM fuel}} \right) 10^3$

$$\text{m}^3/\text{kg fuel}$$

$$= \frac{8.3143 \times 353}{1.013 \times 10^5}\left(\frac{15.28}{46}\right) 10^3$$

$$= 9.62 \text{ m}^3/\text{kg fuel}$$

Products at 150 °C and 1.013 bar. Since the saturation temperature of steam at 1.013 bar is 100 °C, the saturation temperature of steam at its partial pressure in the products is substantially lower than this. Hence, the product water at 150 °C is completely vaporised. Thus from the combustion equation

$$\Sigma \text{ mol gaseous products} = 16.28 \text{ mol/mol fuel}$$

$$\text{Volume of products} = \frac{8.3143T}{p}\left(\frac{\Sigma \text{ mol gaseous products}}{\text{RMM fuel}}\right) 10^3$$

$$= \frac{8.3143 \times 423}{1.013 \times 10^5}\left(\frac{16.28}{46}\right) 10^3$$

$$= 12.29 \text{ m}^3/\text{kg fuel}$$

Products at 15 °C and 1.013 bar. From saturated water and steam tables, partial pressure of saturated steam at product temperature $= p_{steam} = 0.01704$ bar. Thus

$$n_{steam} = \left(\frac{p_{steam}}{p - p_{steam}}\right)(\Sigma \text{ mol dry gaseous products})$$

$$= \left(\frac{0.01704}{1.013 - 0.01704}\right) 13.28$$

$$= 0.2272 \text{ mol/mol fuel}$$

Hence

$$\text{water condensed} = \frac{18(n_2 - n_{steam})}{\text{RMM fuel}}$$

$$= \frac{18(3 - 0.2272)}{46}$$

$$= 1.085 \text{ kg/kg fuel}$$

At dewpoint for products at 1.013 bar

$$p_{steam} = p\left(\frac{n_2}{\Sigma \text{ mol wet gaseous products}}\right)$$

$$= 1.013\left(\frac{3}{16.28}\right)$$

$$= 0.1867 \text{ bar}$$

From saturated water and steam tables

dew point = saturation temperature corresponding to p_{steam}

= 58.6 °C

5 PROPORTIONS OF HOT PRODUCTS

As indicated on p.11, the thermal energy contained in combustion products at temperatures above about 1800 K is sufficient to cause instability, giving rise to dissociation back towards the reactant forms of carbon monoxide and hydrogen, together with other species. At a given high temperature, a condition of dynamic equilibrium can be envisaged at which the rates of oxidation and dissociation are exactly equal so that the product composition remains constant. The following treatment shows how the product composition may be calculated at the given temperature.

5.1 KINETIC EQUILIBRIUM

Development of the kinetic theory of gases permits the calculation of chemical reaction rates in both forward (combustion) and reverse (dissociation) directions. Reaction rates are found *by experiment* to be directly proportional to the instantaneous concentrations, *raised to some power,* of the reacting materials. Thus, in a reversible reaction such as

$$A + B \rightleftharpoons C$$

the instantaneous rate of forward reaction after a given time period might be found by experiment to be in the form

$$\text{forward reaction rate} \propto [A]\,[B]$$

$$= k_F\,[A]\,[B]$$

where $[X]$ = instantaneous molar concentration of reactant X and k_F = rate constant for the forward reaction. The powers to which these concentrations are raised are, in this case, both unity. Similarly, the reverse reaction may well be found by experiment to be given in the form

$$\text{reverse reaction rate} = k_R\,[C]$$

where k_R = rate constant for the reverse reaction. Again, the power is unity.

At the dynamic equilibrium condition for any given temperature, the two rates are equal, hence

$$k_F\,[A]\,[B] = k_R\,[C]$$

and $$\frac{[C]}{[A][B]} = \frac{k_F}{k_R} = K'$$

where K' is known as the *concentration equilibrium constant* for the given reaction. In what follows, reaction products are located in the numerator of the equilibrium constant expressions.

Since, from Avogadro's principle, molar concentrations of gases are proportional to partial pressures

p_X = partial pressure of material X

= total pressure $\left(\dfrac{\text{mol } X}{\text{total mol}}\right)$

Hence, this equilibrium condition can also be represented by

$$\frac{p_C}{p_A p_B} = K$$

where K is known as the *partial-pressure equilibrium constant* for the given reaction. These are the values normally tabulated in the literature (see table 3);[4,5] hence the Avogadro correction of (pressure × molar ratio) is necessary to determine the molar concentrations, except in those cases where the sums of the indices of the numerator and denominator are the same.

With hydrocarbon fuels generally, dissociation occurs from water to hydrogen, and with the high-temperature stage of carbon dioxide to carbon monoxide, thus

$$H_2 + 0.5O_2 \rightleftharpoons H_2O$$

and $$CO + 0.5O_2 \rightleftharpoons CO_2$$

leading to the following expressions for equilibrium

$$\frac{p_{H_2O}}{p_{H_2}\left(p_{O_2}\right)^{\frac{1}{2}}} = K_{H_2O} = \frac{n_{H_2O}}{n_{H_2}\left(p n_{O_2}/n_T\right)^{\frac{1}{2}}} \tag{15}$$

and $$\frac{p_{CO_2}}{p_{CO}\left(p_{O_2}\right)^{\frac{1}{2}}} = K_{CO_2} = \frac{n_{CO_2}}{n_{CO}\left(p n_{O_2}/n_T\right)^{\frac{1}{2}}} \tag{16}$$

[see example 9]

where n_T = total mol present. Since both reactions are occurring together, the half mole of oxygen produced by dissociation of the mole of carbon dioxide may be considered as the oxygen required by the mole of hydrogen. Combination of the two combustion equations then gives

$$CO_2 + H_2 \rightleftharpoons CO + H_2O$$

which is the water-gas reaction, leading to the water-gas reaction equilibrium constant

34

$$\frac{P_{CO} \; P_{H_2O}}{P_{CO_2} \; P_{H_2}} = K_{WG} = \frac{n_{CO} \; n_{H_2O}}{n_{CO_2} \; n_{H_2}}$$

$$= K'_{WG} \text{(in this case)}$$

(17)

[see example 10]

Unless the pressure is very low, further dissociation to radicals and atomic species is unlikely to be extensive at the relatively low temperatures of combustion of hydrocarbons with air, owing to the massive proportion of diluent nitrogen present.

5.2 EQUILIBRIUM PRODUCT COMPOSITION IN HYDROCARBON/AIR COMBUSTION

At any given temperature, the use of published values of partial-pressure equilibrium constants permits the derivation of the relative molar quantities of reactants and products co-existing in equilibrium. For convenience in what follows, all reactants and products are assumed to be in the *gaseous* phase.

5.2.1 Fuel-rich Mixtures Dissociating to Carbon Monoxide and Hydrogen Only

This case is an intermediate one between the non-dissociated cases of chapter 4 and that of general dissociation in the following section. This occurs by not making the assumption that the more reactive hydrogen will be fully oxidised in preference to the carbon, but recognising dissociation only to the extent of sharing the limited oxygen available between carbon dioxide and carbon monoxide. It requires values of the water-gas reaction equilibrium constant for the selected temperatures. The combustion equation therefore contains five unknowns, as follows

$$C_xH_y + m(O_2 + 3.76N_2) = n_1CO_2 + n_2H_2O + n_3CO + n_4H_2 + n_6N_2$$

For given values of x, y and m, the following five equations permit solution.

carbon balance: $n_1 + n_3 = x$

hydrogen balance: $n_2 + n_4 = y/2$

oxygen balance: $n_1 + 0.5n_2 + 0.5n_3 = m$

nitrogen balance: $n_6 = 3.76m$

From water-gas reaction

$$K'_{WG} = \frac{n_3n_2}{n_1n_4} = K_{WG}$$

Since both linear and non-linear equations are involved, solution entails iteration, as in the following procedure.

35

(1) At given temperature, read K_{WG} from table 3.

(2) Assume value of n_1 (slightly below the stoichiometric value).

(3) Evaluate $n_3 = x - n_1$; $n_2 = 2m - 2n_1 - n_3$; and $n_4 = y/2 - n_2$.

(4) Evaluate $(n_3 n_2 / n_1 n_4)$ and compare with K_{WG}.

(5) Repeat from (2) until equality.

5.2.2 General Mixture Dissociating to Carbon Monoxide, Hydrogen and Oxygen

In the more general and realistic case where the fuel and air burn and dissociate to carbon monoxide, free hydrogen *and* oxygen, the general combustion equation, incorporating six unknowns, applies to all mixture ratios

$$C_x H_y + m(O_2 + 3.76N_2)$$

$$= n_1 CO_2 + n_2 H_2 O + n_3 CO + n_4 H_2 + n_5 O_2 + n_6 N_2$$

In addition to the three equations provided by molar balances of carbon, hydrogen and nitrogen, the oxygen molar-balance equation is amended to

$$n_1 + 0.5n_2 + 0.5n_3 + n_5 = m$$

and two further equations result from the dissociation of carbon dioxide and water, as follows

$$K_{CO_2} = \frac{n_1}{n_3 (pn_5/n_T)^{\frac{1}{2}}}$$

and $\quad K_{H_2O} = \dfrac{n_2}{n_4 (pn_5/n_T)^{\frac{1}{2}}}$

where p = total pressure of product mixture and n_T = total mol products present.

Again, solution entails iteration, as in the following procedure.

(1) At given temperature, read K_{CO_2} and K_{H_2O} from table 3.

(2) Assume value of n_5/n_T, and evaluate $(pn_5/n_T)^{\frac{1}{2}}$.

(3) Evaluate $n_1/n_3 = (pn_5/n_T)^{\frac{1}{2}} K_{CO_2}$ and $n_2/n_4 = (pn_5/n_T)^{\frac{1}{2}} K_{H_2O}$.

(4) Evaluate $n_3 = x/(1 + n_1/n_3)$, and $n_1 = x - n_3$,

36

also $n_4 = (y/2)/(1 + n_2/n_4)$, and $n_2 = y/2 - n_4$.

(5) Evaluate $n_5 = [2m - x - (n_1 + n_2)]/2$.

(6) Evaluate $n_T = n_5/(n_5/n_T)$, and compare with $\Sigma n = 3.76m + \Sigma n_j$, where $j = 1$ to 5.

(7) Repeat from (2) until equality.

The concentrations of carbon dioxide and water only are shown in figure 8 for stoichiometric combustion of members of the various

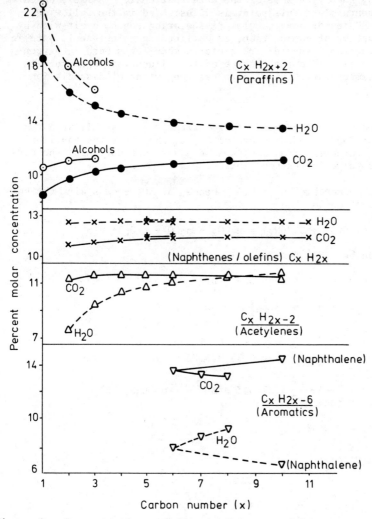

Figure 8   Concentrations of CO_2 and H_2O at stoichiometric combustion temperature, $T_{p\ ad}$ (ref. 1)

hydrocarbon series considered. These indicate a reversal in the relative proportions of water and carbon dioxide as the series range from paraffins to aromatics. (This point is particularly important in relation to combustion temperature levels - see chapter 7.)

[see example 11]

5.3 EXAMPLES

It is apparent that the full calculation of dissociated product composition can be a fairly lengthy process, hence problems of this type appearing in examination papers tend to be foreshortened in some way by supplying some of the data that could, in fact, be calculated eventually. This point is illustrated in the following examples. Cases of more complex dissociation can be dealt with by various systems of computation, as outlined in reference 1. Atomic and other unstable species are expected when fuels burn in oxygen rather than air since the absence of the diluent nitrogen permits a higher temperature, which increases the extent of dissociation.

Example 9

In the combustion of carbon monoxide with 50% excess air at a temperature of 2500 K and pressure of 3 bar, show that the equilibrium volumetric concentration of oxygen is 6.98%, and determine the corresponding value for carbon dioxide.

For carbon monoxide $m_s = 0.5$. Hence, at 50% excess air, the combustion equation is

$$CO + 0.75(O_2 + 3.76N_2) = n_1 CO_2 + n_3 CO + n_5 O_2 + n_6 N_2$$

carbon balance: $n_1 + n_3 = 1$

oxygen balance: $n_1 + 0.5n_3 + n_5 = 0.5 + 0.75 = 1.25$

thus $n_5 = 1.25 - 0.5(n_1 + n_3) - 0.5n_1$

$\qquad = 0.75 - 0.5n_1$

nitrogen balance: $n_6 = 0.75 \times 3.76 = 2.82$

$$K_{CO_2} = \frac{n_1}{n_3(pn_5/n_T)^{\frac{1}{2}}} = 27.543 \text{ atm}^{-\frac{1}{2}} \quad \text{(from table 3)}$$

Given $n_5/n_T = 0.0698$

$$n_1 = (pn_5/n_T)^{\frac{1}{2}} K_{CO_2} n_3$$

$$= \left(\frac{3 \times 0.0698}{1.01325}\right)^{\frac{1}{2}} 27.543n_3$$

$$= 12.521n_3$$

TABLE 3 EQUILIBRIUM CONSTANTS AND ALGEBRAIC TOTAL ENTHALPY CHANGES FOR GRAPHITE AND GASES (Derived from JANAF Tables[4])

Temp. (K)	Partial-pressure Equilibrium Constants			$\Delta H^* = (\Delta H_T + \Delta H_f^o)$ for compounds kJ/mol			$\Delta H^* = \Delta H_T$ for elements kJ/mol				Temp. (K)
	K_{CO_2} (atm$^-$)	K_{H_2O} (atm$^-$)	K_{WG}	CO_2	H_2O	CO	$C(gr)$	H_2	O_2	N_2	
298.15	1.1641×10^{45}	11.169×10^{39}	9.3756×10^{-6}	-393.522	-241.827	-110.529	0	0	0	0	298.15
300	575.44×10^{42}	6.1094×10^{39}	10.617×10^{-6}	-393.455	-241.764	-110.474	0.017	0.054	0.054	0.054	300
500	10.593×10^{24}	76.913×10^{21}	7.2611×10^{-3}	-385.208	-234.906	-104.600	2.381	5.883	6.088	5.912	500
1000	16.634×10^{9}	11.535×10^{9}	0.69343	-360.117	-215.848	-88.843	11.816	20.686	22.707	21.460	1000
1500	207.01×10^{3}	530.88×10^{3}	2.5644	-331.808	-193.732	-71.680	23.230	36.267	40.610	38.405	1500
2000	765.60	3.467×10^{3}	4.5290	-302.072	-169.138	-53.790	35.321	52.932	59.199	56.141	2000
2100	345.94	1.6866×10^{3}	4.8753	-296.022	-163.996	-50.154	37.321	56.379	62.986	59.748	2100
2200	168.27	874.98	5.2000	-289.947	-158.791	-46.509	40.250	59.860	66.802	63.371	2200
2300	87.097	480.84	5.5208	-283.851	-153.532	-42.853	42.727	63.371	70.634	67.007	2300
2400	47.753	277.43	5.8076	-277.734	-148.222	-39.183	45.216	66.915	74.492	70.651	2400
2500	27.543	167.49	6.0814	-271.596	-142.863	-35.505	47.710	70.492	78.375	74.312	2500
2700	10.351	68.077	6.5766	-259.266	-132.014	-28.121	52.727	77.718	86.199	81.659	2700
3000	3.0549	22.029	7.2111	-240.659	-115.466	-16.987	60.300	88.743	98.098	92.738	3000

Note At 298.15 K, ΔH^* for compounds = ΔH_f^o; ΔH_f^o H_2O(L) = -285.7 kJ/mol

39

thus $n_3 = 1/13.521 = 0.0739$

and $n_1 = 1 - n_3 = 0.9261$

$n_5 = 0.75 - 0.4631 = 0.2869$

Therefore

$n_T = 4.1069$

Thus volumetric concentration of $O_2 = 100 \left[\dfrac{0.2869}{4.1069}\right] = 6.98\%$

as assumed, and

volumetric concentration of $CO_2 = 100 \left[\dfrac{0.9261}{4.1069}\right] = 22.55\%$

Example 10

The equilibrium constant for the water-gas reaction: $CO_2 + H_2 \rightleftharpoons CO + H_2O$ is given as 4.8753 at 2100 K. Determine the volumetric concentration of carbon monoxide at this temperature.

Let the·molar fraction of carbon dioxide consumed be q, then the equilibrium combustion equation on a molar basis is

$$(1 - q)CO_2 + (1 - q)H_2 \rightleftharpoons qCO + qH_2O$$

$$K_{WG} = \frac{n_{CO}\, n_{H_2O}}{n_{CO_2}\, n_{H_2}} = \frac{q^2}{(1-q)^2} = 4.8753$$

Thus $3.8753q^2 - 9.7506q + 4.8753 = 0$

and $q = \dfrac{9.7506 \pm \sqrt{(95.0742 - 75.573)}}{7.7506}$

$= 1.83$ or 0.69

Since $(1 - 1.83)$ is negative

$q = 0.69$

Since total mol at any instant, including equilibrium condition, is 2

equilibrium volumetric concentration of CO at 2100 K $= 100 \dfrac{(0.69)}{2}$

$= 34.5\%$

Example 11

Determine the equilibrium product composition for a stoichiometric mixture of iso-octane and air maintained at 2300 K.

From section 5.2.2, the dissociated stoichiometric combustion equation on a molar basis is

$$i\text{-}C_8H_{18} + m_s(O_2 + 3.76N_2)$$

$$= n_1CO_2 + n_2H_2O + n_3CO + n_4H_2 + n_5O_2 + n_6N_2$$

and $m_s = x + y/4 = 12.5$

carbon balance: $n_1 + n_3 = x = 8$

hydrogen balance: $n_2 + n_4 = y/2 = 9$

oxygen balance: $n_1 + 0.5n_2 + 0.5n_3 + n_5 = m_s = 12.5$

thus $n_5 = \dfrac{25 - 8 - (n_1 + n_2)}{2}$

$$= 8.5 - \frac{(n_1 + n_2)}{2}$$

nitrogen balance: $n_6 = 3.76m_s = 3.76 \times 12.5 = 47$

Total molar products $= \Sigma n = 64 + n_5$

Following the procedure in section 5.2.2

(1) For given temperature of 2300 K, $K_{CO_2} = 87.097$, and $K_{H_2O} = 480.84 \text{ atm}^{-\frac{1}{2}}$.

(2) Assume $(n_5/n_T) = 0.01$, then $(pn_5/n_T)^{\frac{1}{2}} = 0.1$.

(3) $n_1/n_3 = (0.1)87.097 = 8.710$, and $n_2/n_4 = (0.1)480.84 = 48.084$.

(4) $n_3 = 8/(1 + 8.710) = 0.824$, and $n_1 = 8 - n_3 = 7.176$;

$n_4 = 9/(1 + 48.084) = 0.183$, and $n_2 = 9 - n_4 = 8.817$.

(5) $n_5 = 8.5 - (7.176 + 8.817)/2 = 0.504$.

(6) $n_T = n_5/(n_5/n_T) = 0.504/0.01 = 50.4$, whereas $\Sigma n = 64.504$

hence $n_T \neq \Sigma n$.

By trial and error, $n_5/n_T = 0.00843$ gives $n_T = \Sigma n = 64.55$, with the following values: $n_1 = 7.111$; $n_2 = 8.801$; $n_3 = 0.889$; $n_4 = 0.199$; $n_5 = 0.544$ and n_6 remains as 47.

6   COMBUSTION ENERGIES

Thermodynamically, a body of combustion products in equilibrium may
be classed as a system, since it is a 'fixed quantity of matter
enclosed by a boundary defining a region in space'. Conservation of
energy demands that any energy transfer across the boundary incurs a
corresponding change in the total stock of energy of the system itself. Energy can transfer across the boundary in the forms of work
W and/or heat Q, and the sign convention is such that work flow outwards and heat flow inwards are both classed as positive. Hence the
total outwards flow of energy = (W - Q), in view of the sign convention, and this corresponds to a loss of equal magnitude in the in-
trinsic energy stock of the system. The following expressions apply
respectively to systems that are stationary, and to those that are
flowing but with negligible changes in potential (height) and kinetic (velocity) energies.

From non-flow energy equation

$$W - Q = - \Delta U$$

$$= \text{change in internal energy of system}$$

From reduced steady-flow energy equation

$$W - Q = - \Delta H$$

$$= \text{change in enthalpy of system}$$

where

enthalpy $H = U + pV$

= the sum of the internal energy and the flow work
entailed initially in forcing the system into the
defined region of volume V against the constant
back pressure p[6]

The energy stock available for transfer across the boundary is
derived, of course, from the prior combustion, and in the two special
cases shown below where the work term is zero, the above expressions
may be simplified to

non-flow heat transfer *at constant volume* = n-f Q_V = ΔU (18)

= internal energy of combustion

and steady-flow heat transfer
at constant pressure = s-f Q_p = ΔH (19)

42

= enthalpy of combustion

This aspect of the subject, concerned with transfer of reaction energy in the form of heat only, is known as thermochemistry, and processes involving inward and outward transfer of heat are known as *endothermic* and *exothermic* respectively. Although thermochemistry is concerned with *changes* in energy, a condition of 25 °C (298.15 K) and 1 atmosphere is selected as datum, and all values related to this are described as standard and indicated by the superscript ^{o}. Thermodynamically, the heat flows resulting from combustion can be represented as areas on the T-s diagram, as shown in figure 9.

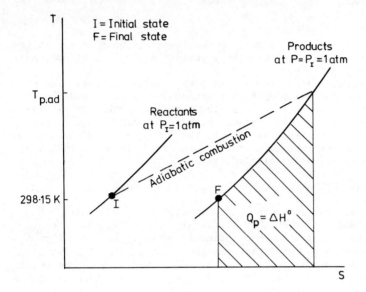

Corresponding figure with slightly steeper gradients for curves of reactants and products at V_I giving $Q_v = \Delta U^o$ for combustion at constant volume.

Figure 9   Representation of heat transfer in constant pressure combustion

A body of reactants in equilibrium can also be classed as a system, on the same basis, and the relationship between changes in H and U as the system changes chemically from one type to another at equal initial and final temperatures is found as follows

$$\Delta H = (H)_P - (H)_R$$

where subscripts P and R refer to products and reactants respectively. Thus

$$\Delta H = (U + pV)_P - (U + pV)_R$$

$$= \Delta U + (pV)_P - (pV)_R$$

Assuming all the gases concerned approximate to ideal gases, the equation of state applies, thus

$$pV = nR_0T \quad \text{as in section 4.6}$$

and $\boxed{\Delta H = \Delta U + R_0T\Delta n}$ (20)

where $\Delta n = (n_P - n_R)$ = change in number of *gaseous* moles owing to the reaction.

Enthalpy has tended to be used for illustrative purposes in the following treatment, but similar arguments apply to internal energy.

Since reaction processes may involve changes in phase, it is necessary for the phase of each reactant and product to be specified. Each material is therefore qualified with one of the symbols (s), (1) or (g) representing solid, liquid or gaseous phase respectively. In the case of solid elemental carbon, the symbol (gr) is used to indicate that the carbon exists in its standard state as graphite, and not as one of its other allotropes.

6.1 STANDARD ENERGY OF FORMATION

The fact that reactants and products exist with some degree of stability indicates that they lie within energy troughs, and are unable to react until further energy is provided to bring them over an energy peak into some other trough. In the case of a fuel molecule, the parent elements - typically $C(gr)$ and $H_2(g)$ - can be imagined existing stably in their energy troughs with no incentive to react. With the addition of a certain quantity of 'atomisation' energy, however, these elements will reach an energy peak and be converted to gaseous atomic $C(g)$ and $H(g)$ respectively by overcoming their bonding forces. These gaseous atoms can then create new bonds with each other to form a stable molecule of hydrocarbon C_xH_y, releasing a certain quantity of 'dissociation' energy (the energy required to dissociate the molecule back to carbon and hydrogen atoms) as the molecule falls into its appropriate energy trough. The net energy change in this formation reaction is termed the *energy of formation*, and is given by

energy of formation = atomisation energy - dissociation energy

In terms of enthalpy, and with initial and final conditions as standard, this appears in the following terms

standard enthalpy of formation = $\Delta H_f^o = \Sigma\Delta H_a - \Sigma D(X-Y)$

as shown in figure 10, where ΔH_a = enthalpy of atomisation required to dissociate the elemental molecular bond, and D(X-Y) = bond dissociation energy required to dissociate the bond X-Y located within a given type of molecule.

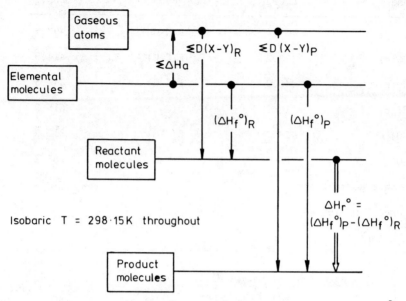

Figure 10 Derivation of standard enthalpies of formation (ΔH_f^o) and reaction (ΔH_r^o)

For convenience, alternative use is made of the empirical bond energy, E(X-Y), which is the mean value of D(X-Y) for a number of such bonds in different locations within molecules. With the water molecule, for example, D(H-OH) for the first bond broken is 497.5 kJ/mol, and D(H-O) for the remaining bond is 428.7 kJ/mol, giving E(H-O) as the average value 463.1 kJ/mol. Representative values of atomisation and dissociation quantities are given in table 4, and of enthalpies of formation in table 5 (at the end of the book) and in figure 11.

In many cases, ΔH_f^o is seen to be negative, indicating that the fuel or product molecule is more stable than its component elemental molecules and is therefore unlikely to revert spontaneously back to them. If any other enthalpy is released as heat during the formation, this also appears as a negative quantity, and gives greater stability to the molecule. Examples of this are the resonance enthalpy of certain molecules that exist as the resultant of a number of slightly differing structures (as with benzene),[1] and also the latent enthalpy absorbed or released due to a change of phase. Consequently, the more complete expression is given by

$$\Delta H_f^o = \Sigma \Delta H_a - \Sigma D(X-Y) - \Delta H_{RESONANCE} \pm \Delta H_{LATENT} \qquad (21)$$

TABLE 4   THERMOCHEMICAL BOND ENERGIES (kJ/mol)
(Derived mainly from references 3 and 5)

Bond	Energy	Bond	Energy
$\Delta H_a H_2(g)$	435.4	D(H-OH)	497.5
$\Delta H_a O_2(g)$	498.2	E(C-H)	414.5
$\Delta H_a N_2(g)$	946.2	E(C-C)	347.5
$\Delta H_a C(gr)$	717.2	E(C=C)	615.5
D(H-O)	428.7	E(C≡C)	812.2
E(H-O)	463.1	E(C-O)	351.7

Resonance Corrections

Benzene, C_6H_6 150.4

Naphthalene, $C_{10}H_8$ 255.4

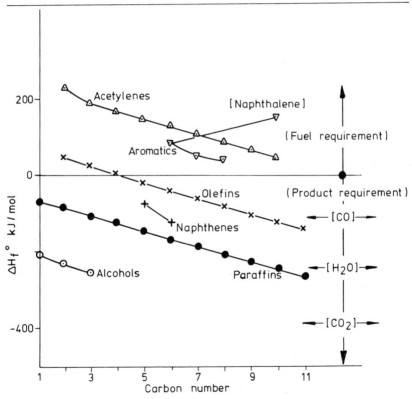

Figure 11   Standard enthalpies of formation of hydrocarbons - gaseous phase (derived from refs 5 and 7)

In the case of such elemental molecules as H_2, O_2, etc., the enthalpy of formation is, of course, zero since $\Delta H_a = D(X-Y)$ as H_2 atomises to 2H and then recombines to H_2, and so on.

[see example 12]

On the above basis, no formation reaction is possible until the full enthalpies of atomisation have been supplied to the elements. In fact, since bond breaking and remaking are occurring together during a transitional stage, reaction may be initiated by the supply of an 'activation energy', E, considerably lower than the full sum of the atomisation enthalpies.

6.2 STANDARD ENERGY OF REACTION

Whereas the energy of formation is seen to relate only to the combination of elemental molecules into a single compound molecule, the generic term 'energy of reaction' covers chemical reactions between elements, compounds or any other species into any number of molecules, atoms or any other species. As before, a certain quantity of energy is found necessary to trigger the atoms of the reactants into rearranging themselves as products, which generally fall into lower energy troughs. The net energy change in this reaction is termed the *energy of reaction* and is given by

energy of reaction = (energy of formation)$_P$

- (energy of formation)$_R$

The standard energy of reaction results when the initial and final conditions are standard and all the materials involved exist in their standard phase. In terms of enthalpy, therefore

standard enthalpy of reaction = $\Delta H_r^o = (\Delta H_f^o)_P - (\Delta H_f^o)_R$

thus $\boxed{\Delta H_r^o = \Sigma n_j (\Delta H_f^o)_j - \Sigma m_i (\Delta H_f^o)_i}$ (22)

as shown in figure 10.

[see example 13]

Values of ΔH_r^o are included in table 5 and plotted in figure 12a.[7] It is helpful at this point to note again that this standard enthalpy of reaction is equal to the transfer of heat owing to the reaction under steady-flow conditions, which are not only isobaric (constant pressure) but also have initial and final temperatures equal.

The above quantity is known as the *standard enthalpy of combustion* when a stoichiometric mixture of fuel (as *gas, liquid* or *solid* depending on its boiling and melting points in relation to standard conditions) and *gaseous* oxygen react to *gaseous* carbon dioxide and *liquid* water only at 298.15 K and 1 atm. These values are negative for a combustion reaction, indicating a release of combustion en-thalpy with the production of products more stable than their parent reactants.

[see example 14]

In engineering terms, heat release is viewed in the positive sense, and mass is a more convenient basis for expressing energy quantities, therefore

$$\text{reaction enthalpy per unit mass fuel} = \frac{-\Delta H_r^o}{\text{RMM fuel}}$$

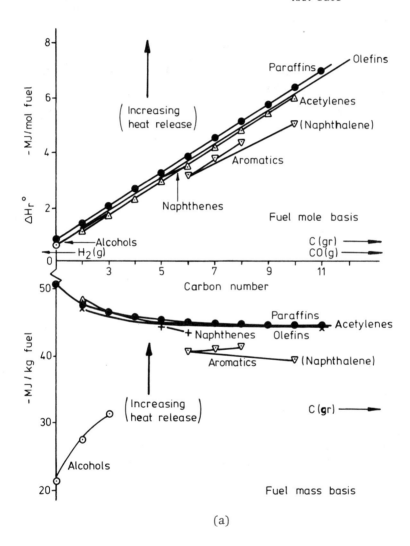

(a)

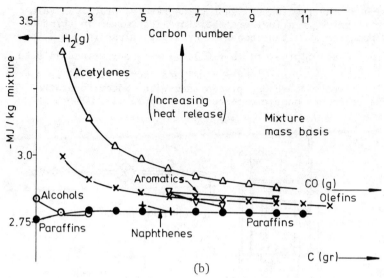

(b)

Figure 12 Standard enthalpies of reaction of hydrocarbons - reactants and products in gaseous phase: (a) fuel basis (derived from refs 7 and 13), (b) stoichiometric fuel/air mixture basis (ref. 1)

Hence the curves in figure 12 have been inverted, and plotted in both types of unit. In some instances, the energy content of a unit mass of fuel/oxidant mixture is of greater significance than that of the fuel alone, consequently

$$\text{reaction enthalpy per unit mass of stoichiometric fuel/air mixture} = \frac{-\Delta H_r^o}{(\text{RMM fuel}) + 137.9 m_s}$$

$$= \frac{-\Delta H_r^o}{1 + (A/F)_s}$$

These values are plotted in figure 12b and show little variation even for substantial differences in ΔH_r^o of fuel, largely because ΔH_r^o and m_s [and thus $(A/F)_s$] vary directly, so that additional air is available to absorb additional heat, and vice versa.

6.3 CALORIFIC VALUE

In engineering practice, wide use is made of non-flow combustion at constant volume in the spark-ignition piston engine, and of steady-flow combustion at constant pressure in the furnace, gas turbine, ramjet, rocket and a variety of chemical and industrial processes. Consequently, expressions 18 and 19 are adopted for the determination of energy release in terms of the outward heat transfer resulting from combustion and subsequent cooling to the (nominal) initial

49

temperature of the reactants.   In the laboratory, practical values
of calorific value of a fuel are therefore determined in terms of
the heat transfer at either non-flow constant volume (n-f Q_V), or

steady-flow constant pressure (s-f Q_p).   The former method is used

for solid and liquid fuels, where a weighed sample of fuel is placed
in a stainless-steel bomb and pressurised with (water-saturated)
oxygen gas to ensure complete combustion (figure 13).   The bomb

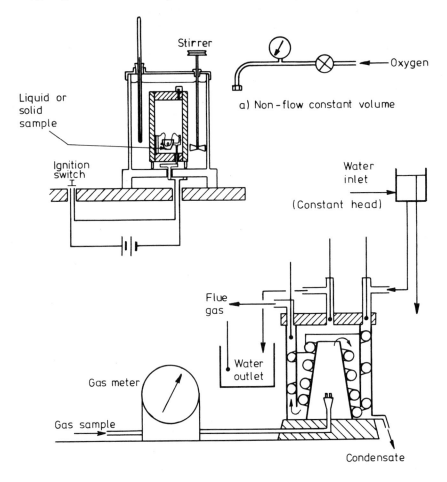

b) Steady - flow constant pressure

Figure 13   Fuel calorimeters: (a) non-flow constant volume, (b)
            steady-flow constant pressure

forms part of a water calorimeter, so that the heat absorbed is the
product of the effective heat capacity and term ΔT_c, where effective

heat capacity is the total energy change per degree K of all the materials making up the calorimeter, including its content of cooling water, and ΔT_c is the corrected temperature rise derived from the test. The instrument is calibrated by an initial determination of effective heat capacity using a fuel of known calorific value (usually benzoic acid), so that determination of the heat absorbed depends solely on measurement of ΔT. This usually does not exceed about 4 °C, and the various corrections allow for heat content of ignition wire and fuel capsule (for the more volatile fuels), together with the formation of nitric and sulphuric acids.[8]

In the 'isothermal' version of the calorimeter, a correction is also made for the heat transfer to the environment during the test. Temperature readings are taken for periods of 5 minutes immediately before and after the combustion, and the mean gradient adopted for the period (Δt) of the combustion itself, to correct the observed temperature rise ΔT_o, as follows

$$\Delta T_c = \Delta T_o - \frac{\Delta t}{2} \text{ (initial temp gradient + final temp gradient)}$$

In the 'adiabatic' version, the calorimeter is fitted with a water jacket, which is maintained electrically at the same temperature as the calorimeter itself, hence no cooling correction is needed. With due allowances made for these small heat transfers, therefore, the heat absorbed by the calorimeter is equal to that released by combustion. Division by the original mass of the fuel sample thus gives the gross calorific value at constant volume

$$GCV_V = \frac{\text{(effective heat capacity)}\,\Delta T_c - \text{energy additions}}{\text{sample mass}} \qquad (23)$$

expressed in units MJ/kg. The two points to note are

(1) Since $\Delta T \neq 0$, and neither the initial nor final temperature is necessarily standard, the resulting calorific value differs slightly from $-\Delta U_r^o$.

(2) The low temperature of the products indicates that the water produced by combustion is almost completely condensed, hence the result is the 'gross' or 'higher' calorific value, incorporating the latent heat of vaporisation of the condensed water. In an engine application, the products must leave the combustion chamber as hot gases (following the second law of thermodynamics) taking the vaporisation heat of the $H_2O(g)$ with them hence the 'net' or 'lower' calorific value is more appropriate, and is determined by subtracting the latent heat of vaporisation from the 'gross' value, as for example

NCV = GCV - 0.212(H) MJ/kg

where H = mass percent of hydrogen in the sample. Alternatively, the value of u_{fg} at 298 K (= 2.30 MJ/kg H_2O) may be used.

[see example 15]

51

For gaseous fuels, the steady-flow constant pressure method is used, with a continuous flow of fuel burning in air within a water-cooled calorimeter. The cooling water flow rate is adjusted to give a slight, measurable rise in temperature, and the gross calorific value calculated in terms of s-f Q_p from this temperature rise, the heat capacity of the water mass flow, and the volumetric flow rate of the fuel gas. As before, the reactant and product temperatures are not equal to 298.15 K, or to each other, hence the resulting calorific value differs slightly from $-\Delta H_r^o$. Allowing for energy remaining in the flue gases, therefore, the gross calorific value at constant pressure is given by

$$GCV_p = \frac{\text{(heat capacity of water) } \Delta T \text{ of water}}{\text{sample volume}} + \frac{\text{flue gas}}{\text{correction}} \qquad (24)$$

[see example 16]

The value of h_{fg} at 298 K (= 2.44 MJ/kg H_2O) may be used for the vaporisation conversion to NCV_p. Calorific values for representative industrial and commercial fuels are plotted in figure 14.

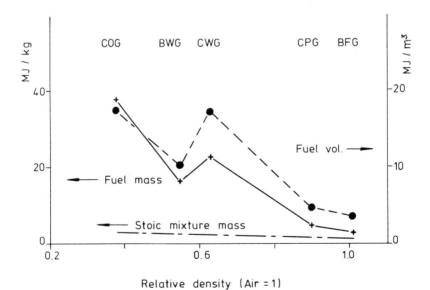

a) Gaseous industrial fuels

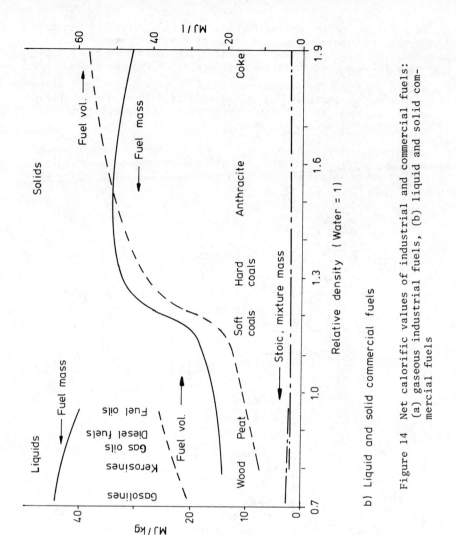

Figure 14 Net calorific values of industrial and commercial fuels:
(a) gaseous industrial fuels, (b) liquid and solid commercial fuels

b) Liquid and solid commercial fuels

6.4 MAXIMUM USEFUL WORK

For work transfer applications in engines, in contrast to heat transfer in furnaces, it is sometimes necessary to find the maximum useful
work that can be transferred from the hot products resulting from
combustion. This is given by the thermodynamic concept of 'availability', which can be defined as the maximum work, less the unusable
work expended in the environment, done by a system as it degrades to
a 'dead' state of equilibrium with its environment. For this work
to be a maximum, both the work and heat transfers must be reversible,
and entropy must not be allowed to rise. Theoretically, a direct
work transfer can be imagined that is isentropic, that is, both re-

versible and adiabatic, extending until the temperature of the product system is reduced from T_1 to that of the environment T_0. This
is likely to bring the pressure of the system down from p_1 to some
temporary level p_x, which is lower than that of the environment p_0.

The second stage of the process therefore consists of a compression
to p_0 carried out isothermally at T_0, resulting in a loss in entropy
of the system equal to the gain in entropy of the environment
(= Q/T_0), that is, a constant entropy over all.

In thermodynamics, work transfer and heat transfer are represented
by the area of the p-V diagram, and the T-S diagram, respectively,
and the above two-stage process is illustrated for the non-flow case
in figure 15, for which

$$\text{n-f } W_{max} = - \Delta U + Q$$

$$= - \Delta U + T_0 \Delta S$$

since $Q = T_0 \Delta S$ for the isothermal compression at T_0.

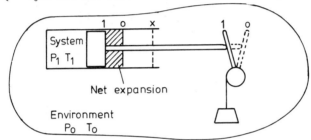

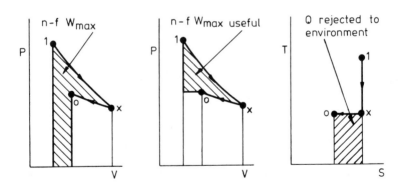

1 to x = isentropic

x to o = isothermal

Area projected from 1, x, o to ordinate represents s-f W_{max}

Figure 15   Maximum useful work from combustion products

In general, the expansion and compression will not be equal, and will result in a net change in volume. If this is positive, some of the work evolved will be consumed as expansion work against the environmental pressure p_0 and this can serve no useful purpose. Consequently, the *non-flow maximum useful work* available externally beyond the requirements of the system and its environment is given by

$$\text{n-f } W_{\text{max useful}} = \text{n-f } W_{\text{max}} - \text{expansion work}$$

$$= (U_1 - U_0) + T_0(S_0 - S_1) - p_0(V_0 - V_1)$$

$$= (U_1 + p_0V_1 - T_0S_1) - (U_0 + p_0V_0 - T_0S_0)$$

thus
$$\boxed{\begin{aligned} \text{n-f } W_{\text{max useful}} &= A_1 - A_0 \\ &= \text{non-flow availability} \end{aligned}}$$

(25)

where $A = (U + p_0V - T_0S) = $ non-flow availability function. Should the system change from state 1 to state 2 without reaching the equilibrium state 0, the maximum useful work available $= A_1 - A_2$.

If the system of combustion gases at state 1 now *flows* through the environment at state 0, the expression for maximum work is modified by the inclusion of the net flow work associated with the system entering and leaving the environment. Since this net flow work contains the expansion work, all the work is useful, hence

$$\text{s-f } W_{\text{max}} = \text{s-f } W_{\text{max useful}}$$

$$= (U_1 - U_0) + T_0(S_0 - S_1) + (p_1V_1 - p_0V_0)$$

$$= (U_1 + p_1V_1 - T_0S_1) - (U_0 + p_0V_0 - T_0S_0)$$

$$= (H_1 - T_0S_1) - (H_0 - T_0S_0)$$

thus
$$\boxed{\begin{aligned} \text{s-f } W_{\text{max useful}} &= B_1 - B_0 \\ &= \text{steady-flow availability} \end{aligned}}$$

(26)

where $B = (H - T_0S) = $ steady-flow availability function. Both A and B are composite properties depending on the state of both the system and its environment.[6]

[see example 17]

6.5 EXAMPLES

In general, enthalpies have been used for illustrative purposes, but one example is given using internal energy to show how the derivation can be made.

Example 12

Calculate the standard enthalpy of formation, in kJ/mol, of the following materials, using the bond energy method: (a) gaseous benzene, given resonance enthalpy = 150.4 kJ/mol, (b) liquid water, given enthalpy of vaporisation = 44.0 kJ/mol.

(a) The formation reaction of gaseous benzene is

$$6C(gr) + 3H_2(g) \longrightarrow 6C(g) + 6H(g) \longrightarrow C_6H_6(g)$$

For thermochemical purposes the structure of the benzene molecule is represented as follows

thus $\Delta H_f^o \ C_6H_6(g) = \Sigma \Delta H_a - \Sigma E(X-Y) - \Delta H_{RESONANCE}$

$$= [6(717.2) + 3(435.4)] - [3(347.5) + 3(615.5)$$

$$+ \ 6(414.5)] - 150.4$$

$$= (4303.2 + 1306.2) - (1042.5 + 1846.5 + 2487.0)$$

$$- \ 150.4$$

$$= 83.0 \ kJ/mol$$

(b) The formation reaction for liquid water is

$$H_2(g) + 0.5O_2(g) \longrightarrow 2H(g) + O(g) \longrightarrow H_2O(l)$$

thus $\Delta H_f^o \ H_2O(l) = \Sigma \Delta H_a - \Sigma E(X-Y) - \Delta H_{LATENT}$

$$= [435.4 + 0.5(498.2)] - 2(463.1) - 44.0$$

$$= 684.5 - 926.2 - 44.0$$

$$= - \ 285.7 \ kJ/mol$$

Example 13

Calculate the standard volumetric energy of reaction in MJ/kg fuel for gaseous benzene reacting to gaseous carbon dioxide and liquid water, given $\Delta H_f^o \ CO_2(g) = -94.054$ kcal/mol using required values from example 12.

By inspection, the stoichiometric combustion equation on a molar basis is

$$C_6H_6(g) + 7.5O_2(g) = 6CO_2(g) + 3H_2O(l)$$

thus $\Delta H_r^o \ C_6H_6(g) = \Sigma n_j (\Delta H_f^o)_j - \Sigma m_i (\Delta H_f^o)_i$

$$= [6(-94.054 \times 4.184) + 3(-285.7)] - (83.0 + 0)$$

$$= [6(-393.522) -857.1)] - 83.0$$

$$= -3301.2 \text{ kJ/mol}$$

and $\Delta U_r^o \ C_6H_6(g) = \Delta H_r^o - R_0 T^o \Delta n$

$$= -3301.2 - \frac{8.3143}{1000} \times 298.15(6 - 8.5)$$

$$= -3295.0 \text{ kJ/mol}$$

$$= \frac{-3295.0}{\text{RMM fuel}} = \frac{-3295.0}{78} = -42.24 \text{ kJ/g(or MJ/kg) fuel}$$

Example 14

Calculate the standard enthalpy of combustion of octane, C_8H_{18}, in MJ/kg fuel and in MJ/kg stoichiometric mixture with air, given $\Delta H_f^o \ C_8H_{18}(l) = -259.5$ kJ/mol, and using required values from examples 11 and 13.

From example 11, $m_s = 12.5$, thus stoichiometric combustion equation on a molar basis is

$$C_8H_{18}(l) + 12.5O_2(g) = 8CO_2(g) + 9H_2O(l)$$

ignoring the nitrogen, therefore

$$\Delta H_c^o \ C_8H_{18}(l) = \Sigma n_j (\Delta H_f^o)_j - \Sigma m_i (\Delta H_f^o)_i$$

$$= [8(-393.522) + 9(-285.7)] - (-259.5 + 0)$$

$$= (-3148.2 - 2571.3) + 259.5$$

$$= -5460.0 \text{ kJ/mol fuel}$$

$$= \frac{-5460.0}{\text{RMM fuel}} = \frac{-5460.0}{114} = -47.89 \text{ MJ/kg fuel}$$

$$= \frac{-5460.0}{\text{RMM fuel} + 137.9 m_s} = \frac{-5460.0}{114 + 137.9(12.5)}$$

$$= -2.97 \text{ MJ/kg stoichiometric mixture}$$

57

Example 15

Determine the net calorific value in MJ/kg of a sample of kerosine given the C/H mass ratio of 6.1/1, and the following test data.

Mass of fuel sample = 0.6081 g
Mass of ignition wire consumed = 0.0148 g
Calorific value of Ni-Cr ignition wire = 1403 J/g
Water equivalent of calorimeter = 740 g
Water added to calorimeter = 1100 g
Initial temperature change during 5 minutes = 15.980 to 16020 °C
Final temperature change during 5 minutes = 19.648 to 19.624 °C
Duration of combustion heating = 2 minutes 45 seconds

Effective heat capacity = (water equivalent + mass water added) 4.1868

$$= (740 + 1100)4.1868$$

$$= 7703.712 \text{ J/K}$$

$$\Delta T_c = \Delta T_o - \frac{\Delta t}{2} \text{ (initial temp. gradient + final temp. gradient)}$$

$$= (19.648 - 16.020) - \frac{2.75}{2} \left(\frac{16.020 - 15.980}{5} + \frac{19.624 - 19.648}{5} \right)$$

$$= 3.628 - \frac{1.375}{5} (0.040 - 0.024)$$

$$= 3.624 \text{ °C}$$

Gross calorific value

$$= \frac{(\text{effective heat capacity})\Delta T_c - (\text{mass wire consumed} \times \text{CV wire})}{\text{sample mass}}$$

$$= \frac{(7703.712)3.624 - (0.0148 \times 1403)}{0.6081}$$

$$= 45876 \text{ kJ/kg}$$

$$= 45.88 \text{ MJ/kg approximately}$$

Net calorific value = gross calorific value $- \frac{0.212(100)}{1 + \text{C/H}}$

$$= 45.88 - 21.2/7.1$$

$$= 42.99 \text{ MJ/kg approximately}$$

Note As an engineering quantity, calorific value is expressed as positive, whereas the thermochemical reaction enthalpy is expressed as negative.

Example 16

Determine the gross and net calorific values of a gaseous fuel burnt in a constant pressure gas calorimeter given the following test results.

Atmospheric pressure = 101.19 kPa
Atmospheric temperature = 295 K
Volume of gaseous sample consumed = 0.01 m^3
Volume of water heated = 4014 cm^3
Temperature of measured water = 308 K
At 308 K, temperature correction of water volume = 0.4%
Initial temperature of water = 293.10 K
Final temperature of water = 315.62 K
Flue gas temperature = 299 K
Volumetric correction of flue-gas enthalpy = 8.27 kJ/m^3 K
Volume of condensate = 8.7 cm^3 (= 8.7 g approximately)
Latent enthalpy of steam (h_{fg}) at 288 K = 2.47 kJ/g

$$\text{Standardised volume of gaseous sample} = 0.01 \left(\frac{101.19}{101.325}\right)\left(\frac{288}{295}\right)$$

$$= 0.00975 \ m^3$$

$$\text{Temperature-corrected volume of heated water} = 0.996 \times 4014$$

$$= 3998 \ cm^3$$

Thus heat capacity of water = 3.998×4.1868

$$= 16.72 \ kJ/K$$

Enthalpy correction of flue gas = $8.27(299 - 295)$

$$= 33.08 \ kJ/m^3$$

Gross calorific value

$$= \frac{\text{(heat capacity of water)}\Delta T \text{ of water}}{\text{sample volume}} + \text{flue-gas correction}$$

$$= \frac{16.72(315.62 - 293.10)}{0.00975} + 33.08$$

$$= 38695 \ kJ/m^3$$

$$= 38.7 \ MJ/m^3 \text{ approximately}$$

$$\text{Net calorific value} = 38.7 - \frac{8.7 \times 2.46}{0.00975 \times 10^3}$$

$$= 36.5 \ MJ/m^3 \text{ approximately}$$

Example 17

Find the maximum, and maximum useful, specific work (kJ/kg) that could be derived from combustion products that are (a) stationary, and (b) flowing, in an environment under the following conditions.

	p bar	T K	v m^3/kg	u kJ/kg	s kJ/kg K
Products (1)	7	1000	0.41	760.0	7.4
Environment (0)	1	298.15	0.83	289.0	6.7

(a) n-f w_{max} = $(u_1 - u_0) + T_0(s_0 - s_1)$

$\qquad\qquad$ = $(760.0 - 289.0) + 298.15(6.7 - 7.4)$

$\qquad\qquad$ = $471.0 - 208.7$

$\qquad\qquad$ = 262.3 kJ/kg

$\qquad$ n-f $w_{max\ useful}$ = n-f w_{max} - $p_0(v_0 - v_1)$

$\qquad\qquad\qquad$ = $262.3 - 100(0.83 - 0.41)$,
since 1 bar = 100 kPa

$\qquad\qquad\qquad$ = 220.3 kJ/kg

(b) s-f w_{max} = s-f $w_{max\ useful}$ = $(h_1 - T_0s_1) - (h_0 - T_0s_0)$

$\qquad\qquad$ = $(u_1 - u_0) + (p_1v_1 - p_0v_0) + T_0(s_0 - s_1)$

$\qquad\qquad$ = n-f w_{max} + $(p_1v_1 - p_0v_0)$

$\qquad\qquad$ = $262.3 + 100\ [(7 \times 0.41) - 0.83]$

$\qquad\qquad$ = 466.3 kJ/kg

7 COMBUSTION TEMPERATURES

The molar balance data of chapter 5 permitted the derivation of
equilibrium product composition at any selected temperature, which
might be sustained artificially by the extraction or supply of
energy in some form. In combination with the energy-balance data
of chapter 6, it now becomes possible to determine the temperature
achieved naturally by the combustion products under any specified
conditions. Assuming no external addition from the environment,
the maximum value of this temperature is reached when no energy
losses occur through work or heat transfers to the environment,
that is, under isochoric adiabatic conditions in the non-flow case,
and isobaric adiabatic conditions in steady flow. Once again, in
view of the many applications of constant-pressure steady flow in
furnaces, continuous-flow engines and chemical processing generally,
the following treatment is based on enthalpy, with the symbol $T_{p\ ad}$
representing isobaric adiabatic combustion temperature. However,
comparable arguments apply to internal energy and $T_{V\ ad}$ for the isochoric adiabatic combustion temperature in such applications as the
spark-ignition piston engine.

As a first step to the determination of these combustion (flame)
temperatures, it is necessary to be able to quantify the energy absorbed by the combustion products due to the reaction, and also by
the reactants in the event of any preheating.

7.1 SENSIBLE ENERGY

Both ΔU and ΔH are seen to result in transfers of energy in the form
of heat alone in the two sets of conditions outlined on p.42, the
driving force for heat transfer being a difference in temperature.
In fact, both U and H are found to be functions of temperature alone
for ideal gases, and approximately so for real gases. The thermal,
or sensible, energy content can therefore be related directly to the
level of temperature obtaining and, in terms of enthalpy, values of
absolute enthalpy $(H_T - H_0)$, or of enthalpy above standard
$(H_T - H_{298.15})$, are tabulated against temperature T in the literature
[4, 5, 9, 10, 11] for various product materials together with simple
hydrocarbon fuels such as methane and acetylene. Consequently, the
rise in sensible enthalpy of a material from standard 298.15 K to
some temperature T is given by

$$\Delta H_T = H_T - H_{298.15}$$

$$= (H_T - H_0) - (H_{298.15} - H_0)$$

Values of ΔH_T in kcal/mol have been extracted from the literature

and converted to kJ/mol, as discussed in the following section.
Values of sensible enthalpy for reactants are needed only when the
initial conditions are non-standard, as in the case of pre-heating.
For the more complex fuels, values of sensible enthalpy over the
temperature range ΔT concerned can be obtained by means of expressions in terms of the temperature changes, molar quantities and
molar heat capacities, thus

$$\text{sensible enthalpy of more complex fuel} = \Delta H_T$$

$$= \Delta T m_F \bar{C}_p$$

where m_F = mol of fuel, $\bar{C}_p$ = mean molar heat capacity at constant
pressure over temperature range ΔT. For mixtures of gases, this
can be written as

$$\Delta H_T = \Delta T \Sigma \left[(m_f)_i (\bar{C}_p)_i \right]$$

For approximate calculations, use can be made of standard
(298.15 K) values of C_p^o, ignoring variations with temperature. For
more accurate work, the variation of C_p with temperature can be expressed as

$$C_p = a + bT + cT^2 + \ldots$$

where a, b, etc., are constants. Thus enthalpy absorbed by one mol
of the more complex fuel heated from 298.15 K over range ΔT is given
by

$$\Delta H_T = \int_{298.15}^{T} C_p \, dT = a\Delta T + \frac{b}{2}(T^2 - 298.15^2) + \frac{c}{3}(T^3 - 298.15^3)$$

Values of C_p and/or a, b, c, etc., for representative fuels (in the
gaseous phase) are available in the literature.[3, 12]

7.2   DETERMINATION OF MAXIMUM TEMPERATURE IN STEADY FLOW

In the case of a chemical reaction occurring in steady flow under
constant pressure, the energy of the reaction has been shown to
promote a change in enthalpy, which is usually defined by the quantity of heat transferred in bringing the temperature of the products
down to that of the reactants. Under standard conditions, this
gives the standard enthalpy of reaction as

$$\text{s-f } Q_p \text{ at standard initial and final temperature} = \Delta H_r^o$$

$$= (\Delta H_f^o)_P - (\Delta H_f^o)_R$$

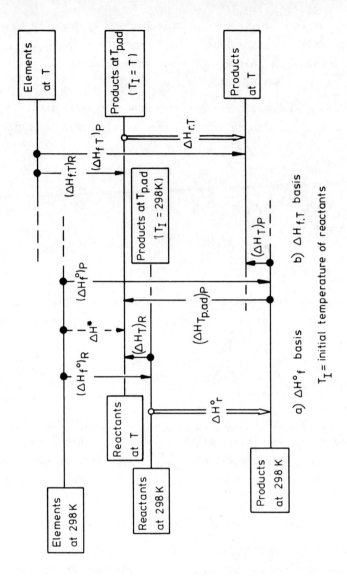

Figure 16 Algebraic total enthalpy changes, ΔH^*

When the initial and final temperatures are equal but are *non*-standard, figure 16b shows that

$$\text{s-f } Q_{p\ T} = \Delta H_{r\ T}$$

$$= (\Delta H_{f\ T})_P - (\Delta H_{f\ T})_R$$

$$= \Sigma n_j (\Delta H_{f\ T})_j - \Sigma m_i (\Delta H_{f\ T})_i$$

63

comparable to the standard expression in equation 22, where $\Delta H_{r\ T}$ = enthalpy of reaction at initial and final temperature T, $\Delta H_{f\ T}$ = enthalpy of formation at initial and final temperature T, n_j = mol of product j and m_i = mol of reactant i.

Values of $\Delta H_{f\ T}$ are obtainable directly from the JANAF tables.[4] Figure 16a shows $\Delta H_{r\ T}$ also to be obtainable by algebraic summation, as follows

$$\text{s-f } Q_{p\ T} = \Delta H_{r\ T}$$

$$= \Delta H_r^o + (\Delta H_T)_P - (\Delta H_T)_R$$

$$= (\Delta H_T + \Delta H_f^o)_P - (\Delta H_T + \Delta H_f^o)_R$$

$$= (\Delta H^*)_P - (\Delta H^*)_R$$

where ΔH_T = rise in sensible enthalpy from 298.15 to initial (and final) temperature T K and ΔH^* = algebraic total enthalpy change from 298.15 to T K = $(\Delta H_T + \Delta H_f^o)$. Thus

$$\text{s-f } Q_{p\ T} = \Sigma n_j (\Delta H^*)_j - \Sigma m_i (\Delta H^*)_i$$

It is noteworthy that the term 'net' enthalpy change might appear more appropriate from the nature of the diagram, but this could lead to confusion since a negative value of ΔH_f^o is tabulated as such, and its insertion into a *net* equation already containing a negative sign could inadvertently lead to arithmetic summing of the two enthalpy quantities. This problem did not arise with ΔH_a and D(X-Y) since the latter is defined, and tabulated, in the positive sense.

Since ΔH_f^o is zero for the elemental reactants and products, their algebraic total enthalpy changes at non-standard temperatures consist of the sensible enthalpy changes alone, that is, $\Delta H^* = \Delta H_T$ alone at temperature T, and = zero at 298.15 K, for elements. Values of ΔH^* for graphite, hydrogen, oxygen, nitrogen and gaseous product molecules are included in table 3 in units of kJ/mol.

In the more general case where the final temperature is higher than the initial temperature T_I, the reaction energy is partly absorbed by the products, and only partly transferred as heat, the proportions depending on the conditions. For this final temperature to be a maximum, all the reaction energy must be absorbed by the

products, and none transferred from the system. Consequently, for
reactants in steady flow, combustion must be not only isobaric for
zero work transfer, but also adiabatic for zero heat transfer. Hence
the conditions giving rise to the maximum combustion temperature in
steady flow $(T_{p\ ad})$ represent a thermal balance between energy released and absorbed. In figure 16, the algebraic total enthalpy
changes are illustrated as common to both reactants and products,
hence

$$\text{s-f } Q_{p\ ad} = (\Delta H_{T_{p\ ad}} + \Delta H_f^o)_P - (\Delta H_{T_I} + \Delta H_f^o)_R$$

that is

$$\text{s-f } Q_{p\ ad} = \Sigma n_j (\Delta H^*)_j - \Sigma m_i (\Delta H^*)_i \qquad (27)$$

$$= \text{zero}$$

In the event that calculations are made at some isobaric product
temperature T_p that is *not* the adiabatic value, the two algebraic
total enthalpy changes will not be equal, and the transfer of the
heat s-f Q_p will have some finite positive or negative value. The
method of solution for $T_{p\ ad}$, therefore, is to determine the dissociation product composition at two selected bracketing temperatures
(using the technique given in section 5.2.2) and to test for s-f Q_p
= 0 in each case, the required value of $T_{p\ ad}$, together with the related product composition, being found by linear interpolation between the values of s-f Q_p at the bracketing temperatures. The following procedure permits solution.

(1) At first value of temperature T_p, determine product composition,
     as in section 5.2.2.

(2) Read all values of algebraic total enthalpy changes from table
     3, using T_I for reactants, and T_p for products.

(3) Weight each value of algebraic total enthalpy change from (2)
     with appropriate molar concentrations from (1), and check for
     s-f Q_p = 0.

(4) Repeat from (1) with bracketing value of temperature T_p.

(5) Determine $T_{p\ ad}$ (and molar concentrations if required) by linear
     interpolation.

Calculations over a wide range of T_p show linear interpolation to
be entirely acceptable. A tabulated method of solution is strongly
recommended, on the following lines.

65

Material		First value of T_p		Second value of T_p
	mol	ΔH^*	$mol(\Delta H^*)$	
(Products)				
CO_2	n_1			
etc.				Similar treatment
(Reactants)				
Fuel	1	$- (-\Delta H^*)$		
O_2	m	$-\Delta H^*$		
N_2	3.76m			
s-f Q_p			=	

[see examples 18 and 19]

7.3 INFLUENCE OF FUEL TYPE AND OPERATING PARAMETERS

Since the maximum combustion temperature is a function of both the
energy released by the reactants and the energy absorbed by the products, it is dependent over all on values of enthalpy of formation,
air/fuel ratio, initial temperature of reactants, reaction pressure,
proportions of resulting products and their levels of heat capacity.
In particular, it is noteworthy that the energy released by hydrogen
is about four times that released by carbon, whereas the specific
heat capacity of water is about twice that of carbon dioxide and carbon monoxide. Consequently, as shown in figures 1 and 8, reduction
in hydrogen content from paraffins to aromatics would be expected to
show comparable reductions in energy release and also in heat capacity. The over-all relationship between $T_{p\ ad}$ and hydrocarbon molecular structure is therefore not immediately predictable.

Values of $T_{p\ ad}$ have been calculated manually for stoichiometric
mixtures of hydrocarbons in air, but with all reactants and products
in the gaseous phase (approximately equivalent to the gross calorific value condition) in the manner outlined in section 7.2, for the
simple cases of dissociation to carbon monoxide, hydrogen and oxygen
only. The results presented in figure 17 show the heat capacity of
the water product to have an overriding influence, with the levels
of $T_{p\ ad}$ increasing broadly from paraffins to aromatics, roughly
comparable to the variation in C/H mass ratio. They also show the

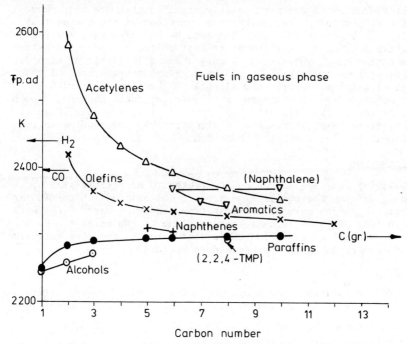

Figure 17   Dissociated stoichiometric combustion temperatures,
$T_{p\ ad}$ (ref. 1)

highest $T_{p\ ad}$ to be given by acetylene at 2583 K, and the average
level of $T_{p\ ad}$ for the fuels shown to be about 2340 K.   Further calculations show the following effects.

(1) The hypothetical non-dissociated values of $T_{p\ ad}$ would be about
    120 K higher.

(2) More extensive dissociation to atomic oxygen and hydrogen, and
    to radicals OH and NO, reduce $T_{p\ ad}$ by about 23 K.

(3) Fuels in the *liquid* phase show reductions in $T_{p\ ad}$ of the follow-
    ing order: paraffins 47 K, aliphatic alcohols 60 K, naphthenes
    10 K, olefins 11 K, and aromatics 58 K.

Increases of ΔT_I in the initial temperature of the reactants are
found to increase $T_{p\ ad}$ by about $0.5\Delta T_I$ only, owing to the increased
dissociation at the higher temperatures.   Increases in reaction pressure increase $T_{p\ ad}$ owing to the restricted dissociation.   Variation
in mixture strength indicates that $T_{p\ ad}$ reaches a maximum value at
about 5% fuel enrichment.[13]

For the commercial fuel blends, the flame temperatures of the

lighter hydrocarbons are particularly significant since they represent the temperatures reached by those portions of the blends that
are first to vaporise and burn, and also by the materials resulting
from thermal cracking of the heavier fuel molecules under the action
of combustion heat. Consequently many commercial blends of liquid,
solid or gaseous fuels exhibit adiabatic flame temperatures of about
2300 K. Flame temperatures are slightly higher for fuels rich in
carbon monoxide and/or hydrogen, and lower for fuels containing such
inerts as carbon dioxide, nitrogen and ash. In heat-transfer applications, flame temperatures may be significantly lower owing to the
intentional loss of energy by radiation.

7.4 EXAMPLES

The following examples have been restricted to manageable size by
providing data on the equilibrium product compositions at particular
temperatures, which make up the most time-consuming part of the work.
These include one example using oxygen alone as the oxidant, the products incorporating atomic and OH species owing to the more extensive
dissociation at the higher temperatures reached.

Example 18

Determine the maximum flame temperature for a stoichiometric mixture
of gaseous iso-octane and air flowing at standard initial conditions,
using molar data from example 11, and given ΔH_f^o for i-C_8H_{18}(g) =
-224.287 kJ/mol, with the following equilibrium product composition
at 2200 K for one mole of fuel[12]

$$7.412CO_2 + 8.865H_2O + 0.588CO + 0.135H_2 + 0.361O_2 + 47N_2$$

$$\text{s-f } Q_p = \Sigma n_j (\Delta H^*)_j - \Sigma m_i (\Delta H^*)_i$$

T_p K		2200			2300	
Material	mol	ΔH^*	mol(ΔH^*)	mol	ΔH^*	mol(ΔH^*)
CO_2	7.412	-289.947	-2149.087	7.111	-283.851	-2018.464
H_2O	8.865	-158.791	-1407.682	8.801	-153.532	-1351.235
CO	0.588	-46.509	-27.347	0.889	-42.853	-38.096
H_2	0.135	59.860	8.081	0.199	63.371	12.611
O_2	0.361	66.802	24.116	0.544	70.634	38.425
N_2	47	63.371	2978.437	47	67.007	3149.329
Fuel	1	$-(-\Delta H^*)$	224.287	1	$-(-\Delta H^*)$	224.287
s-f Q_p			-349.195			16.857

68

Thus $T_{p\ ad}$ = 2200 + 100 $\left(\dfrac{349.195}{349.195 + 16.857}\right)$

$\qquad\qquad$ = 2295 K

Example 19

A stoichiometric gaseous mixture of hydrogen and oxygen flowing at standard conditions is found to give the following combustion equation

$$H_2 + 0.50_2 = n_2H_2O + n_4H_2 + n_5O_2 + n_8H + n_9O + n_{10}OH$$

Determine the isobaric temperature of combustion when a heat loss of 80 kJ/mol fuel occurs, given the values of n and algebraic total enthalpy changes for each product (below) at temperatures 2400 and 3000 K.

The two reactants, being elemental molecules at standard conditions, have algebraic total enthalpy changes of zero, thus the reactant term in the Q_p equation vanishes, leaving

$$\text{s-f } Q_p = \Sigma n_j (\Delta H^*)_j - (-80)$$

Constructing an enthalpy table (incorporating the values of n_j and ΔH^* given with this question) gives

T_p K	2400			3000		
Product	mol (given)	ΔH^* (given)	$mol(\Delta H^*)$	mol (given)	ΔH^* (given)	$mol(\Delta H^*)$
H_2O	0.959	-148.222	-142.145	0.754	-115.466	-87.061
H_2	0.031	66.915	2.074	0.157	88.743	13.933
O_2	0.119	74.492	8.865	0.053	98.098	5.199
H	0.003	261.676	0.785	0.068	274.148	18.642
O	0.001	293.240	0.293	0.030	305.771	9.173
OH	0.016	107.303	1.717	0.109	129.047	14.066
		- Heat loss	80		- Heat loss	80
s-f Q_p			-48.411			53.952

Thus $T_p = 2400 + 600\left(\dfrac{48.411}{48.411 + 53.952}\right)$

$\qquad = 2683$ K

8 COMBUSTION EFFICIENCIES

Ideally, all the energy released by combustion would be available for transfer across the boundary of the products system. In practice, not all the chemical reaction energy is made available owing to imperfections of mixing, incomplete combustion by local chilling, unusable heat flow through the system boundary, and so on. Consequently, a parameter of combustion efficiency arises, of the following form

$$\frac{\text{combustion}}{\text{efficiency}} = \eta_c = \frac{\text{energy actually released by combustion}}{\text{energy theoretically released by combustion}}$$

Should the actual mixture strength be stoichiometric or fuel weak, the denominator applies to complete combustion and is equal to $(\dot{F} \times NCV)$, where $\dot{F}$ = fuel mass flow rate, and NCV = net calorific value of fuel.

In the furnace application, this energy is required for direct transfer as heat to a prescribed environment (made up of water for steam raising, metal stock for heat treatment, chemical stock for processing, etc.). Consequently a radiant flame is needed, which tends to favour high-density high C/H ratio fuels, and efficiency is based on the combined aspects of combustion and heat transfer to this environment. In the engine application, on the other hand, the energy is required to be retained within the products until it can be transferred as work, the extent of energy conversion being subject to the restraint of the second law of thermodynamics. Since the events within the piston engine cannot easily be separated, combustion efficiency as defined above tends to have an over-all value, made up of combustion and conversion to work. In the continuous-flow engine, however, event duration is equivalent to fluid displacement, and it is feasible to measure the temperature of the working fluid before and immediately after combustion, and so assess the efficiency of the combustion process alone.

8.1 WORK TRANSFER APPLICATIONS IN NON FLOW

The thermodynamic performance of reciprocating-piston engines is assessed against the theoretical Otto cycle in the case of the spark-ignition (S-I) and high-speed compression-ignition (C-I) engines, and against the theoretical Diesel cycle in the case of the low-speed C-I engine (figure 18). Thermodynamic analyses of these theoretical cycles give the following expressions for efficiency.

$$\eta_{Otto} = 1 - (1/r_V)^{\gamma-1}$$

and
$$\eta_{Diesel} = 1 - \left[\frac{\alpha^\gamma - 1}{\gamma(\alpha - 1)}\right] (1/r_V)^{\gamma-1}$$

where r_V = compression ratio = cylinder volume at bottom dead centre/ cylinder volume at top dead centre, α = cut-off ratio (or load ratio) and γ = ratio of specific heat capacities, c_p/c_V = 1.4 approximately, for air.

It is noteworthy that thermal efficiency for the Otto cycle is equivalent to that of the Joule constant-pressure heat-transfer cycle used for the performance assessment of the gas turbine, although this is normally expressed in terms of the pressure ratio r_p, as

$$\eta_{Joule} = 1 - (1/r_p)^{(\gamma-1)/\gamma}$$

since $r_p = (r_V)^\gamma$.

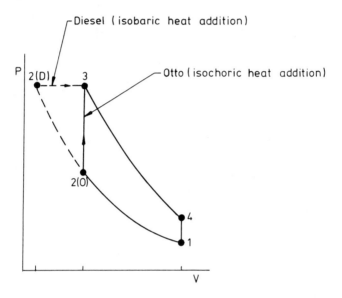

Compression ratio $= r_v = V_1/V_2$

Cut-off ratio (Diesel) $= \alpha = V_3/V_{2(D)}$

Figure 18   Theoretical Otto and Diesel thermodynamic cycles

Curves in figure 19 show the variations of these ideal thermal efficiencies with r_V and/or r_p. Owing to imperfections in mixing, combustion, mixture distribution, and fuel loss through crankcase dilution, actual 'indicated' values of thermal efficiency are lower than the ideal values shown in the figure, and are given by

$$\frac{\text{combustion}}{\text{efficiency}} = \frac{\text{indicated power}}{\text{mass flow fuel} \times \text{net calorific value fuel}}$$

72

$$= \eta_c = \frac{ip}{\dot{F} \; NCV} \qquad (28)$$

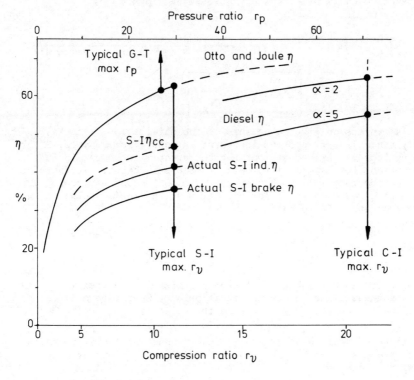

γ = 1.4

S-I = Spark - ignition engine

C-I = Compression - ignition engine

G-T = Gas - turbine engine

Figure 19   Representative values of engine thermal efficiencies

Further losses from friction, pumping, windage, radiation, etc., in transmitting the power to the crankshaft and flywheel give 'brake' values of thermal efficiency, which are again reduced, as shown in the figure.

A more closely defined efficiency of combustion is possible from a conclusion drawn in chapter 6 that the net energy release per unit mass of stoichiometric mixture of fuel with air is remarkably constant at about 2.8 MJ/kg *mixture* over a wide range of hydrocarbon fuels. This corresponds to a net value of about 3.0 MJ/kg *air* in the mixture, which equally tends to be common throughout the range of fuels considered, and can be described as the net calorific value of air (NCV_A).

Provided that the mixture lies on the rich side of stoichiometric (as in the S-I engine), all the air is available for combustion, and the mass flow of air together with its net calorific value provide an effective basis of efficiency assessment. Hence

$$\text{combustion chamber efficiency} = \frac{\text{indicated power}}{\text{mass flow air} \times \text{net calorific value air}}$$

$$= \eta_{cc} = \frac{ip}{\dot{A}\ NCV_A} \tag{29}$$

This expression gives a measure of the utilisation of the air supply since it reflects the increment in power (for example, 3%) resulting from an enrichment, but not the greater increment in fuel consumption (for example 10%) owing to the enrichment. It therefore excludes the fuel losses that appear in a combustion-efficiency determination based on the calorific value of the fuel. Values of η_{cc} range from about 34 to 45%.

[see example 20]

8.2 HEAT TRANSFER APPLICATIONS IN STEADY FLOW

As indicated above, substantial heat transfer is required instantly from a furnace flame, hence the equilibrium temperature achieved by the flame is considerably less than the value of $T_{p\ ad}$. The over-all efficiency of heat release plus transfer to the prescribed environment is then resolved as a mathematical product of the efficiencies of the process of combustion and the subsequent process of heat transfer. Thus

$$\text{boiler (or furnace) efficiency} = \text{combustion efficiency} \times \text{heat transfer efficiency}$$

$$= \frac{\text{rate of energy gain of prescribed environment}}{\text{rate of energy supply}}$$

In this study, combustion efficiency is the major concern, and this can be expressed as

$$\text{combustion efficiency} = \eta_c = \frac{\text{rate of energy release}}{\text{rate of energy supply}}$$

$$= \frac{\text{rate of energy supply} - (\text{rate of energy losses from unburnt CO and C in ash})}{\text{rate of energy supply}}$$

$$= \eta_c = 1 - \frac{\dot{m}_{C/CO}\ (\Delta H_{C/CO_2} - \Delta H_{C/CO}) + \dot{m}_{C/ash}\Delta H_{C/CO_2}}{\dot{F}\ CV_F} \tag{30}$$

74

where $\dot{m}_{C/CO}$ = mass rate of C burning to CO, kg/h, $\dot{m}_{C/ash}$ = mass rate
of C rejected in ash, kg/h, $\dot{F}$ = mass rate of fuel supplied, kg/h,
$\Delta H_{C/CO_2}$ = reaction enthalpy of C burning to CO_2, MJ/kg, $\Delta H_{C/CO}$ =

reaction enthalpy of C burning to CO, MJ/kg and CV_F = calorific value

of fuel, MJ/kg. (*Note* The gross value is frequently used, but the
net value is more appropriate since the flue gases leave the chamber
hot.)

[see example 21]

8.3 WORK TRANSFER APPLICATIONS IN STEADY FLOW

For the constant-pressure steady-flow case, as in the gas turbine or
ramjet combustion chamber, the energy quantities are expressed in
terms of enthalpies as follows

$$\eta_c = \frac{\text{actual gain in enthalpy of fluid flow per second}}{\text{theoretical gain in enthalpy of fluid flow per second}}$$

$$= \frac{\dot{H}_P - (\dot{H}_A + \dot{H}_F)}{\dot{F} \; NCV}$$

Hence the 'energy based' combustion efficiency is given by

$$\boxed{\text{energy } \eta_c = \frac{(\dot{A} + \dot{F})\bar{c}_{p_2}T_{t_2} - (\dot{A}\bar{c}_{p_1}T_{t_1} + \dot{F}\bar{s}T_F)}{\dot{F} \; NCV}} \qquad (31)$$

where $\dot{H}_P$, $\dot{H}_A$ and $\dot{H}_F$ are enthalpy flow rates of products, air and

fuel respectively, T_{t_1} and T_{t_2} are the total-head temperatures of

the air and products respectively, T_F is the fuel inlet temperature,
$\dot{A}$ and $\dot{F}$ are the mass flow rates of air and fuel respectively, NCV
is the net calorific value of the fuel, $\bar{c}_{p_1}$ and $\bar{c}_{p_2}$ are the mean
specific heat capacities of air and products respectively over their
respective ranges of temperature and $\bar{s}$ is the mean specific heat
capacity of fuel from datum to T_F.

If the temperature of the products varies across the outlet duct
of the combustor, it is usual to express T_{t_2} as the mean of a number

of values obtained from a temperature traverse across equal sub-areas
of the outlet cross-section. If mass flow variations occur also, individual temperatures are weighted by corresponding mass flow rates
to give the weighted-mean outlet temperature value of T_{t_2}.

For rapid results in development work, certain approximations may
be made. Ignoring the sensible enthalpy of the fuel, the kinetic
energy of the relatively low-speed entering air, together with dif-

ferences between actual and theoretical values of specific heat capacities of air, fuel and products reduces the energy expression to one
of temperature rises only, as follows

$$\frac{\text{'temperature based'}}{\text{combustion efficiency}} = \frac{\text{actual total-head temperature rise}}{\text{theoretical total-head temperature rise}}$$

thus

$$\text{temperature } \eta_c = \frac{T_{t_2} - T_1}{(T_{t_2}' - T_1)} \tag{32}$$

where T_1 = measured static temperature of inlet air (approximately
equal to the total-head temperature T_{t_1}), T_{t_2} = measured weighted
mean total-head temperature of products and T_{t_2}' = theoretical total-
head temperature of products for complete combustion. Data on mean
specific heat capacities and rises in total-head temperature are
available in various forms in the literature.[14] In the case of combustion temperature rises of about 500 K, temperature efficiencies
are usually found to exceed the corresponding energy values by about
0.05% for each 1% loss of efficiency.

An alternative method of approximation is based on the conversion
of carbon-bearing compounds to carbon dioxide. Individual samples
taken from a traverse are analysed for carbon dioxide by means of an
instrument such as an infrared gas analyser, both before and after
oxidation in a platinum-packed electrically heated furnace to give
net and gross values respectively. The method then depends on the
assumption that the local inefficiency in each area sampled is due
to the missing carbon dioxide, and is equivalent to the energy that
the main fuel (rather than the actual unburnt and partially burnt
constituents) would release in providing this carbon dioxide. Hence
local 'CO_2 based' combustion efficiency is given by

$$\text{local } CO_2 \ \eta_c = \frac{\text{mol C present as CO}}{\text{total mol C present}}$$

that is

$$\text{local } CO_2 \ \eta_c = \frac{\text{net } CO_2}{\text{gross } CO_2} \tag{33}$$

For an over-all value of carbon dioxide efficiency, the net carbon
dioxide concentrations can be weighted for mass flow through each
measuring area to give a weighted mean net carbon dioxide concentration, as with the treatment of T_{t_2}. For fuel of a specified carbon
content, the furnace oxidation can be dispensed with, and the weighted mean net carbon dioxide concentration compared with the calculated gross carbon dioxide concentration for the over-all operating
value of A/F mass ratio.

Under optimum conditions, aero gas-turbine combustors operate at about 98 to 99% efficiency, but this drops appreciably if tests are made at atmospheric pressure.

[see example 22]

8.4 EXAMPLES

These concern the combustion efficiencies of a piston engine, and the various expressions for combustion efficiency used in the furnace and gas-turbine combustor.

Example 20

A spark-ignition reciprocating-piston engine operating with an iso-octane fuel at 10% enrichment develops indicated power given by ip = 1100$\dot{A}$ kW where $\dot{A}$ = mass flow of air in kg/s. Calculate the values of combustion efficiency and combustion chamber efficiency given the net calorific value of iso-octane as 44.74 MJ/kg.

Octane = C_8H_{18}, thus

$$\text{gravimetric } (A/F)_s = \frac{137.9(x + y/4)}{12x + y}$$

$$= \frac{137.9(12.5)}{114}$$

$$= 15.12$$

At 10% enrichment

$$A/F = 15.12/1.1 = 13.75$$

$$\text{Net CV air} = NCV/(A/F)_s = 44.74/15.12 = 2.96 \text{ MJ/kg}$$

Combustion
efficiency = η_c

$$= \frac{\text{indicated power}}{\text{mass flow fuel} \times \text{net calorific value fuel}}$$

$$= \frac{ip}{\dot{F} \text{ NCV}}$$

$$= \frac{1100\dot{A}}{\dot{A}(F/A)NCV}$$

$$= \frac{1100(A/F)}{NCV}$$

$$= 1100 \times 13.75/44740$$

$$= 0.338$$

77

Combustion chamber
efficiency $= \eta_{cc}$

$$= \frac{\text{indicated power}}{\text{mass flow air} \times \text{net calorific value air}}$$

$$= \frac{ip}{\dot{A} \ NCV_A}$$

$$= \frac{1100}{2960}$$

$$= 0.372$$

Example 21

A steam boiler consumes 700 kg/h of coal of net calorific value 35
MJ/kg and carbon content 85% by mass, and produces 90 kg/h of ash
consisting of 25% carbon by mass. The volumetric analysis of the
dry flue gas includes 13.4% carbon dioxide, 1.5% carbon monoxide
and zero hydrogen. Determine the combustion efficiency of the
boiler furnace given the calorific values of carbon to carbon dioxide and carbon to carbon monoxide as 32.8 and 10.1 MJ/kg carbon
respectively. Find also the boiler efficiency for a steam production rate of 5370 kg/h at 15 bar and 350 °C superheat temperature where enthalpy is 3148 kJ/kg.

Mass carbon supplied $= 0.85 \times 700 = 595$ kg/h

Mass carbon rejected in ash $= \dot{m}_{C/ash} = 0.25 \times 90 = 22.5$ kg/h

Thus mass carbon burnt $= 595 - 22.5 = 572.5$ kg/h

$$\left(\frac{CO_2}{CO}\right) \text{ products molar ratio} = \left(\frac{C \longrightarrow CO_2}{C \longrightarrow CO}\right) \text{molar, or mass, ratio}$$

$$= 13.4/1.5$$

$$= 8.93$$

Hence

mass rate of C burning to CO $= \dot{m}_{C/CO} = 572.5/9.93$

$$= 57.65 \text{ kg/h}$$

$$\eta_c = 1 - \frac{\dot{m}_{C/CO}(\Delta H_{C/CO_2} - \Delta H_{C/CO}) + \dot{m}_{C/ash}\Delta H_{C/CO_2}}{\dot{F} \ CV}$$

$$= 1 - \frac{57.65(32.8 - 10.1) + 22.5(32.8)}{700 \times 35}$$

78

$$= 1 - 2046.66/24\ 500$$

$$= 0.916$$

$$\eta_{boiler} = 5370 \times 3.148/24\ 500$$

$$= 0.690$$

Example 22

Determine the combustion efficiencies, on the bases of energy, temperature and carbon dioxide, for a gas-turbine combustor given the following test conditions.

Air mass flow rate = $\dot{A}$ = 0.500 kg/s

Fuel mass flow rate = $\dot{F}$ = 0.00613 kg/s

Air inlet temperature = T_1 = 300 K(= T_{t_1} nearly enough)

Fuel temperature = T_F = 300 K

Weighted mean total-head temperature of products = T_{t_2} = 755 K

Net calorific value of fuel = NCV = 43.12 MJ/kg

Mean specific heat capacity of air at 300 K = $\bar{c}_{p1}$ = 1.001 kJ/kg K

Mean specific heat capacity of fuel at 300 K = $\bar{s}$ = 1.415 kJ/kg K

Mean specific heat capacity of products at 755 K and F/A 0.01226 = $\bar{c}_{p2}$ = 1.035 kJ/kg K

Theoretical total-head temperature rise for T_{t_1} 300 K and F/A 0.01226 = 493 K

Weighted mean net CO_2 = 2.38%

Gross CO_2 for F/A 0.01226 = 2.60%

$$\text{energy } \eta_c = \frac{(\dot{A} + \dot{F})\bar{c}_{p2}T_{t_2} - (\dot{A}\bar{c}_{p1}T_{t_1} + \dot{F}\bar{s}T_F)}{\dot{F}\ \text{NCV}}$$

$$= [(0.50613)1.035 \times 755 - (0.5 \times 1.001 \times 300)$$

$$- (0.00613 \times 1.415 \times 300)]/(0.00613 \times 43210)$$

$$= \frac{395.503 - 150.15 - 2.602}{264.326}$$

$$= 0.918$$

79

$$\text{temperature } \eta_c = \frac{T_{t_2} - T_1}{(T_{t_2}' - T_1)}$$

$$= (755.300)/493$$

$$= 0.923$$

$$CO_2\eta_c = \frac{\text{net } CO_2}{\text{gross } CO_2}$$

$$= 2.38/2.60$$

$$= 0.915$$

9   OXIDATION IN THE FUEL CELL

Although not a process of combustion in the accepted sense of oxidation accompanied by heat and light, stoichiometric oxidation forms
the basis of direct conversion from chemical to electrical energy by
means of the fuel cell.   In contrast to primary and secondary cells,
fuel and oxidant are supplied continuously during operation rather
than being stored within the cell itself;   consequently the electrodes
are not consumed.   Furthermore, the fuel and oxidant do not meet as
such, as is the case in combustion, but are permitted to react only
through the intermediary of some form of ionised molecule.   Consequently the two electrodes are separated by an electrolyte, which
must be a good conductor for ions, but an insulator for electrons
which are required to be stripped from the reactants and to flow
through an external circuit to do work (figure 20).

The fuel cell represents the ultimate in conversion efficiency
since it not only minimises the number of conversion stages, each

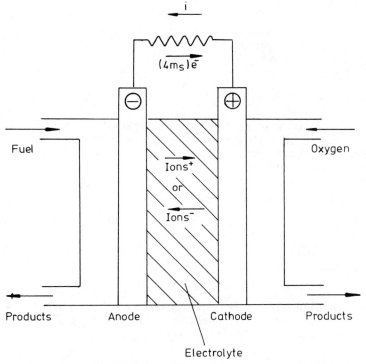

Figure 20   Schematic of fuel cell

with its associated inefficiency, but it is not dependent on high-
temperature heat release, and is therefore freed from the restraint

of the Carnot efficiency of $(1 - T_{min}/T_{max})$. The performance of a fuel cell may be assessed by equating the thermodynamic and the electrical expressions for output work flow. Thermodynamically, the device involves a reactant system in steady flow that provides work transfer at the expense of its stock of enthalpy, associated with a heat transfer, which is generally so minor that it can be considered to be both isothermal and reversible. In terms of the steady-flow energy equation (in its reduced form with negligible changes in potential and kinetic energy)

$$s\text{-}f\ W = -\ \Delta H + Q$$

Since the minor heat transfer is isothermal and reversible

$$s\text{-}f\ W = -\ \Delta H + T\Delta S$$

$$= -\ \Delta G \text{ J/mol fuel}$$

where G = Gibbs free-energy function = $(H - TS)$.

From electrical considerations, on the other hand

$$s\text{-}f\ W = Eit$$

where E = electrical potential developed and i = current flowing during time period t. Thus

$$s\text{-}f\ W = En'e$$

$$= E(N_0 n)e \text{ J/mol fuel}$$

where n' = number of electrons flowing in external circuit/mol fuel, n = number of electrons flowing in external circuit/molecule fuel, N_0 = number of molecules/mol fuel = Avogadro's number = 602.47×10^{21} and e = charge/electron = 0.16021×10^{-18} C. Equating the two expressions for s-f W gives

$$E = \frac{-\Delta G}{N_0 en}$$

$$= \frac{-\Delta G}{96.522 \times 10^3 n} \text{ V}$$

Since oxygen is divalent, with each atom approaching the stability of neon by absorbing two electrons to produce $O^=$, one molecule of oxygen is associated with four electrons, and thus m_s molecules of oxygen with $4m_s$ electrons, where

m_s = stoichiometric molecules O_2/molecule fuel

= stoichiometric mol O_2/mol fuel, as in the stoichiometric combustion equation

Thus $\boxed{n = 4m_s}$ (34)

for any fuel oxidised in a fuel cell, whether or not the ions involved are, in fact, $O^=$ since the over-all reaction is represented by stoichiometric oxidation, even if the ions involved in the electrolyte are OH^-, H_3O^+, H^+, $CO_3^=$ or anything else. Hence

$$E = \frac{-\Delta G}{386.088 \times 10^3 m_s} \text{ V}$$ (35)

The absolute efficiency of the fuel cell is given by the ratio of the change in free energy, which determines the capacity of the reaction to occur, to the change in enthalpy, which should result when the reaction actually does occur. Thus, absolute efficiency is given by

$$\text{absolute } \eta = \frac{-\Delta G}{-\Delta H}$$ (36)

[see example 23]

The difference in these energy quantities, equal to $-\Delta G - (-\Delta H)$, gives rise to the minor heat transfer.

9.1 REACTIONS AT ELECTRODES

If the type of ion involved is known (as determined by the chemical nature of the electrolyte), the direction of ion migration from one electrode to the other is fixed, and the electrochemical reaction at each electrode can be deduced. For example, if the oxidation of a hydrocarbon proceeds via the migration of H^+ ions from anode to cathode, the procedure is as follows.

Over all reaction

$$C_x H_y + m_s O_2 = xCO_2 + \frac{y}{2} H_2O$$

$$m_s = x + y/4$$

$$n = 4(x + y/4)$$

Identifying the reactants and products by $[X]_R$ and $[Y]_P$ respectively and, for convenience, writing the cathode reaction from right to left gives

83

Anode reaction

$$\left[C_xH_y\right]_R + 2xH_2O \longrightarrow 4(x + y/4)e^- + 4(x + y/4)H^+ + \left[xCO_2\right]_P$$

         electrolyte   circuit   electrolyte

Cathode reaction

$$2(x + y/4)H_2O \longleftarrow 4(x + y/4)e^- + 4(x + y/4)H^+ + \left[(x + y/4)O_2\right]_R$$

$$\left[(y/2)H_2O\right]_P$$

In the high-temperature hydrocarbon/oxygen fuel cell, the relevant ion is $CO_3^=$, and its concentration in the electrolyte is maintained by recycling the excess carbon dioxide found at the anode, as follows.

Anode reaction

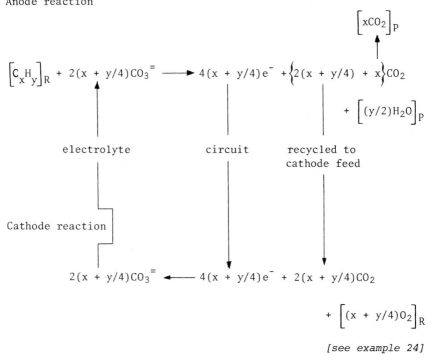

$$\left[xCO_2\right]_P$$

$$\left[C_xH_y\right]_R + 2(x + y/4)CO_3^= \longrightarrow 4(x + y/4)e^- + \left\{2(x + y/4) + x\right\}CO_2$$

$$+ \left[(y/2)H_2O\right]_P$$

electrolyte   circuit   recycled to
                                        cathode feed

Cathode reaction

$$2(x + y/4)CO_3^= \longleftarrow 4(x + y/4)e^- + 2(x + y/4)CO_2$$

$$+ \left[(x + y/4)O_2\right]_R$$

[see example 24]

9.2 EXAMPLES

The type of ion produced and the electrochemical reactions at each electrode cannot be deduced without knowledge of the electrolyte. However, the over-all reaction is then based on the equivalent use of doubly ionised oxygen as the ion, and the performance parameters of electrical potential and absolute efficiency can be obtained directly from values of ΔG and ΔH for the oxidation reaction.

Example 23

Determine the electrical potential developed and the absolute efficiency, for fuel cells using the following fuels with oxygen.

	Fuel	ΔG kJ/mol	ΔH kJ/mol
(a)	H_2	-237.2	-285.8
(b)	C	-394.4	-393.8
(c)	CH_4	-818.5	-891.0

(a) $m_s = 0.5$, thus

$$E = \frac{-\Delta G}{386.088 \times 10^3 m_s} = \frac{237.2 \times 10^3}{386.088 \times 10^3 \times 0.5}$$

$$= 1.23 \text{ V}$$

absolute $\eta = \Delta G / \Delta H = 237.2/285.8 = 83.0\%$

(b) $m_s = 1$, thus

$$E = \frac{394.4 \times 10^3}{386.088 \times 10^3 \times 1} = 1.02 \text{ V}$$

absolute $\eta = 100\%$ since $\Delta G > \Delta H$

(c) $m_s = 2$, thus

$$E = \frac{818.5 \times 10^3}{386.088 \times 10^3 \times 2} = 1.06 \text{ V}$$

absolute $\eta = 818.5/891.0 = 91.9\%$

Example 24

Deduce the electrochemical reactions occurring at the electrodes of the following fuel cells given that the electrolytes are selected to produce the ion types shown.

(a) H_2-O_2 cell with OH^- ion (aqueous hydroxyl electrolyte)

(b) H_2-O_2 cell with H_3O^+ ion (acid electrolyte)

(c) H_2-O_2 cell with H^+ ion (catalytic ion-exchange membrane)

(d) H_2-O_2 cell with $CO_3^=$ ion (fused carbonate electrolyte)

(e) C-O_2 cell with $O^=$ ion

(f) CH_4-O_2 cell with H^+ ion

(a) H_2-O_2, OH^- ion Stoichiometric equation: $H_2 + 0.5O_2 = H_2O$
thus $m_s = 0.5$, and $n = 4(0.5) = 2$.

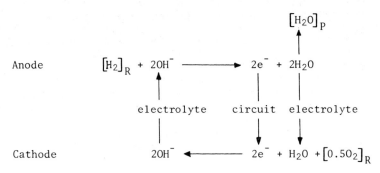

Anode $[H_2]_R + 2OH^- \longrightarrow 2e^- + 2H_2O \longrightarrow [H_2O]_P$

electrolyte   circuit   electrolyte

Cathode $2OH^- \longleftarrow 2e^- + H_2O + [0.5O_2]_R$

(b) H_2-O_2, H_2O^+ ion

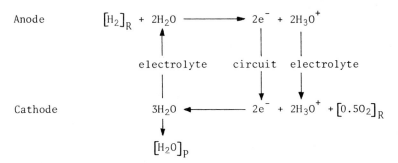

Anode $[H_2]_R + 2H_2O \longrightarrow 2e^- + 2H_3O^+$

electrolyte   circuit   electrolyte

Cathode $3H_2O \longleftarrow 2e^- + 2H_3O^+ + [0.5O_2]_R$

$[H_2O]_P$

(c) H_2-O_2, H^+ ion

Anode $[H_2]_R \longrightarrow 2e^- + 2H^+$

circuit   electrolyte

Cathode $[H_2O]_P \longleftarrow 2e^- + 2H^+ + [0.5O_2]_R$

(d) H_2-O_2, $CO_3^=$ ion

Anode $[H_2]_R + CO_3^= \longrightarrow 2e^- + CO_2 + [H_2O]_P$

electrolyte   circuit   electrolyte

Cathode $CO_3^= \longleftarrow 2e^- + CO_2 + [0.5O_2]_R$

86

(e) $C-O_2$, $O^=$ ion Stoichiometric equation: $C + O_2 = CO_2$ thus $m_s = 1$, and $n = 4$.

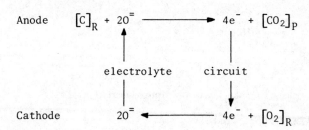

Anode $[C]_R + 2O^= \longrightarrow 4e^- + [CO_2]_P$

electrolyte   circuit

Cathode $2O^= \longleftarrow 4e^- + [O_2]_R$

(f) CH_4-O_2, H^+ ion Stoichiometric equation: $CH_4 + 2O_2 = CO_2 + 2H_2O$ thus $m_s = 2$, and $n = 8$.

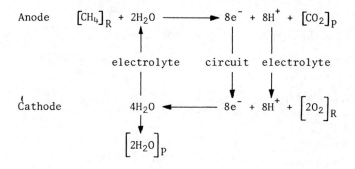

Anode $[CH_4]_R + 2H_2O \longrightarrow 8e^- + 8H^+ + [CO_2]_P$

electrolyte   circuit   electrolyte

Cathode $4H_2O \longleftarrow 8e^- + 8H^+ + [2O_2]_R$

$[2H_2O]_P$

10   SUMMARY

For hydrocarbon fuel C_xH_y

(1)   relative molecular mass = RMM = $12x + y$ approximately

(2)   C/H mass ratio = $12x/y$ approximately

For any fuel-air mixture

(3)   equivalence ratio = $\phi = \dfrac{F/A}{(F/A)_s}$

For any material

(4)   mass material = mol material × RAM or RMM material

For any hydrocarbon fuel, the general combustion equation on a molar basis is

(5) $1(\text{fuel}) + m(O_2 + 3.76N_2) = n_1CO_2 + n_2H_2O + n_3CO + n_4H_2 + n_5O_2$
$$+ n_6N_2$$

For hydrocarbon fuel C_xH_y

(6)   stoichiometric mol O_2/mol fuel = $m_s = x + y/4$

(7)   volumetric $(A/F)_s = 4.76(x + y/4)$

(8)   gravimetric $(A/F)_s = \dfrac{137.9(x + y/4)}{12x + y}$ approximately

For combustion products

(9)   volume % product j = $100 \left(\dfrac{\text{mol j}}{\Sigma \text{mol products}}\right)$

For volumetric fuel blend $(aC_xH_y + bH_2 + cCO + dO_2 + \text{incombustibles})$

(10) $m_s = (x + y/4)a + 0.5b + 0.5c - d$

(11) gravimetric $(A/F)_s$

$$= 137.9 \left[\frac{\Sigma(m_s \text{ 'X'} \times \text{molar fraction 'X'}) - \text{molar fraction } O_2}{\Sigma(\text{RMM fuel component} \times \text{molar fraction fuel component})}\right]$$

approximately, where 'X' is an oxygen-consuming fuel component.

For gravimetric fuel blend ($aC + bH_2 + cO_2 + fS$ + incombustibles)

(12) $M_s = \dfrac{1a}{12} + \dfrac{0.5b}{2} - \dfrac{c}{32} + \dfrac{f}{32}$

(13) gravimetric $(A/F)_s$

$$= 137.9 \left[\Sigma \left(\dfrac{m_s \text{ 'X' } \times \text{ mass fraction 'X'}}{\text{RMM 'X'}} \right) - \dfrac{\text{mass fraction } O_2}{32} \right]$$

approximately, where 'X' is an oxygen-consuming fuel component.

The volume change resulting from combustion, assuming all gases ideal

(14) $\Delta V_C = 8.3143 \left[\dfrac{T_P}{P_P}(\Sigma \text{mol gaseous products}) - \right.$

$\left. \dfrac{T_R}{P_R}(\Sigma \text{mol gaseous reactants}) \right] m^3/\text{mol fuel}$

Partial-pressure equilibrium constants

(15) $K_{H_2O} = \dfrac{n_{H_2O}}{n_{H_2}(pn_{O_2}/n_T)}$

(16) $K_{CO_2} = \dfrac{n_{CO_2}}{n_{CO}(pn_{O_2}/n_T)}$

(17) $K_{WG} = \dfrac{n_{CO}n_{H_2O}}{n_{CO_2}n_{H_2}} = K'_{WG}$

For a thermodynamic system

(18) non-flow heat transfer at constant volume = n-f $Q_V = \Delta U$

(19) steady-flow heat transfer at constant pressure = s-f $Q_p = \Delta H$

In an isothermal chemical reaction

(20) $\Delta H = \Delta U + R_0 T \Delta n$

(21) standard enthalpy of formation = ΔH_f^o

$= \Sigma \Delta H_a - \Sigma D(X-Y) - \Delta H_{RESONANCE} \pm \Delta H_{LATENT}$

(22) standard enthalpy of reaction = ΔH_r^o

$$= \Sigma n_j (\Delta H_f^O)_j - \Sigma m_i (\Delta H_f^O)_i$$

At constant volume

(23) gross calorific value = GCV_V

$$= \frac{(\text{effective heat capacity}) \Delta T_c - \text{energy additions}}{\text{sample mass}}$$

At constant pressure

(24) gross calorific value = GCV_p

$$= \frac{(\text{heat capacity of water}) \Delta T \text{ of water}}{\text{sample volume}} + \text{flue gas correction}$$

For a thermodynamic system

(25) non-flow availability = n-f $W_{\text{max useful}}$ = $A_1 - A_0$ where $A = (U + p_0 V - T_0 S)$

(26) steady-flow availability = s-f $W_{\text{max useful}}$ = $B_1 - B_0$ where $B = (H - T_0 S)$

For combustion in isobaric and adiabatic steady flow, maximum flame temperature is given by

(27) s-f $Q_{p \ ad} = \Sigma n_j (\Delta H^*)_j - \Sigma m_i (\Delta H^*)_i = 0$ where $T_j = T_{p \ ad}$, and $T_i = T_I$

For spark-ignition reciprocating-piston engine

(28) combustion efficiency = η_c = ip/$\dot{F}$ NCV

(29) combustion chamber efficiency = η_{cc} = ip/$\dot{A}$ NCV_A

For steady flow heat transfer

(30) combustion efficiency

$$\eta_c = 1 - \left[\frac{\dot{m}_{C/CO}(\Delta H_{C/CO_2} - \Delta H_{C/CO}) + \dot{m}_{C/ash} \Delta H_{C/CO_2}}{\dot{F} \ CV_F} \right]$$

For steady-flow work transfer

(31) 'energy based' combustion efficiency

$$\text{energy } \eta_c = \frac{(\dot{A} + \dot{F}) \bar{c}_{p_2} T_{t_2} - (\dot{A} \bar{c}_{p_1} T_{t_1} + \dot{F} \bar{s} T_F)}{\dot{F} \ NCV}$$

90

(32) 'temperature based' combustion efficiency

$$\text{temperature } \eta_c = \frac{T_{t_2} - T_1}{(T_{t_2}' - T_1)}$$

(33) local 'CO_2 based' combustion efficiency

$$\text{local } CO_2\eta_c = \frac{\text{net } CO_2}{\text{gross } CO_2}$$

In the fuel cell

(34) number of electrons produced/molecule fuel = $n = 4m_s$

(35) electrical potential developed = $E = \dfrac{-\Delta G}{386.088 \times 10^3 m_s}$ V

(36) absolute efficiency = $\dfrac{-\Delta G}{-\Delta H}$

Physical Quantities

For an ideal gas

$$pV = nR_0T$$

and $pV_M = R_0T$

where n = number of mol of gas, V_M = molar volume = 22.4136 m^3/kmol
at 1 atm and 0 °C and R_0 = universal gas constant = 8.3143 kJ/kmol K.

N_0 = Avogadro's number

= number of molecules/mol fuel

= 602.47×10^{21}

e = charge/electron

= 0.16021×10^{-18} C

II   PROBLEMS

1. A sample of coal has the following gravimetric percentage composition

   carbon 81.2, hydrogen 4.8, oxygen 7.0, ash 3.4, water 3.6

When burnt with excess air, the Orsat analysis of the dry products gives carbon dioxide 10.1%, oxygen 9.5%, with nitrogen as balance, and it is known that some carbon is retained in the ash. Determine the gravimetric air/fuel ratio, and the mass percentage of unburnt carbon.

[16.19;   14.7%]

2. Gaseous hydrocarbon fuel and dry air flow steadily into a combustor, and the products leave with negligible changes in kinetic and potential energy. Given that the volumetric percentage analysis of the wet products is carbon dioxide 8.0, water 10.6, oxygen 6.6, and the remainder nitrogen, determine the gravimetric carbon/hydrogen ratio of the fuel, and the air/fuel ratio of the mixture.

[4.53;   23.42]

3. A furnace designed for fuelling with coal gas at 60% excess air is to be converted to natural gas to give the same power output at the same volumetric supply rate of air. Determine the excess air resulting in the converted condition, together with the volumetric composition of the cooled dry combustion products, assuming complete combustion, given

volumetric percentage composition

   coal gas: CH_4 30, C_2H_4 3.6, CO 8, H_2 52, O_2 0.4, CO_2 2, N_2 4

   natural gas: CH_4 90, CO_2 2, N_2 8

net volumetric calorific values, MJ/m^3

   CH_4 35.8, C_2H_4 58.1, CO 12.6, H_2 10.8

[13.7%;   10.27%;   2.79%;   86.94%]

4. A furnace is fired simultaneously with fuel oil (mass % carbon 88, hydrogen 12) and natural gas (volume % methane 96, carbon dioxide, 2, nitrogen 2). The volumetric percentage analysis of the dry products gives carbon dioxide 10, carbon monoxide 0.6, oxygen 4.2, and the remainder nitrogen. Determine the volume of natural gas, at 101.3 kPa and 288 K, consumed per kilogram of fuel oil.

[3.24 m^3/kg]

5. A combustion test with hydrocarbon fuel $C_{10}H_{22}$ burning with 20% excess air at pressure 1 bar is found to produce carbon dioxide, water, carbon monoxide, hydrogen, oxygen and nitrogen. Determine the gravimetric air/fuel ratio. Find also the molar quantity of each product at 2300 K, given mol oxygen product/mol fuel = 3.46, and partial-pressure equilibrium constant K_{CO_2} = 87.097 $atm^{\frac{1}{2}}$.

[18.06; 9.43; 10.85; 0.57; 0.15; 3.46; 69.94]

6. A hydrocarbon fuel C_xH_y burns at pressure p with 24% excess air at a temperature resulting in volumetric percentage dissociation of 'a' and 'b' in carbon dioxide and water respectively. Derive a formula for the total mol of products per mol fuel, and for the partial-pressure equilibrium constant of the reversible H_2O reaction.

$$\left[(5.90 + a/200)x + (1.73 + b/400)y; \quad \frac{(100 - b)}{b(p)^{\frac{1}{2}}\left[\frac{(96 + 2a)x + (24 + b)y}{(2360 + 2a)x + (692 + b)y}\right]^{\frac{1}{2}}} \right]$$

7. A gaseous mixture consisting of two moles of methane and one of propane is mixed with 10% excess air at 25 °C, and burnt in a constant volume chamber. Determine the pressure at the combustion temperature of 2200 K given that 98.6% of the carbon oxidises completely, and using values of K_{CO_2} and K_{H_2O} from table 3. Find also the initial pressure of the mixture, and the volume of the chamber.

[9.54 bar; 1.27 bar; 0.98 m^3]

8. Determine the percentage of thermal energy in the fuel transferred to the flue gases when methane of net calorific value 35.8 MJ/m^3 is burnt stoichiometrically with air at initial temperature 298 K producing flue gases at 600 K, given the following mean volumetric heat capacities at constant pressure, in kJ/m^3 K, over the range of temperature: carbon dioxide = 1.88, water = 1.56, nitrogen = 1.32.

[12.6%]

9. Determine the enthalpy of reaction of a stoichiometric mixture of acetylene and air at initial and final temperature of 500 K given the algebraic total enthalpy change of acetylene at 500 K as +236.946 kJ/mol.

[-1257.5 kJ/mol]

10. For the gaseous paraffinic fuel propane, C_3H_8, determine

(a) the stoichiometric air/fuel ratio by volume and by mass

(b) the standard enthalpy of formation using data from table 4, and compare with the experimental value of -103.9 kJ/mol

(c) the standard enthalpy of all-gaseous reaction, using the above

93

experimental value of ΔH_f^o

(d) the adiabatic dissociated combustion temperature at constant 1 atm pressure, and the dissociated molar product composition for the stoichiometric mixture with air, using data from table 3, given n_5/n_T = 0.00842 approx. at 2300 K, and 0.00553 approx. at 2200 K

(e) the change in gaseous volume/kg fuel due to combustion at $T_{p\ ad}$

(f) the cooled volumetric percentage composition of the undissociated products, both wet and dry, for the stoichiometric mixture with air

(g) the mass of water condensed/kg fuel when undissociated stoichiometric products are cooled to 15 °C at 1 atm, together with the dewpoint at this pressure.

$$\begin{bmatrix} \text{(a) 23.8, 15.67;} \quad \text{(b) -117.8 kJ/mol;} \quad \text{(c) -2044.0 kJ/mol;} \quad \text{(d) 2295 K,} \\ \text{2.67, 3.91, 0.33, 0.09, 0.21, 18.8;} \quad \text{(e) 97.5 m}^3\text{/kg;} \quad \text{(f) wet: 11.6\%,} \\ \text{72.9\%, 15.5\%,} \quad \text{dry: 13.8\%, 86.2\%;} \quad \text{(g) 1.48 kg/kg, 54.9 °C} \end{bmatrix}$$

TABLE 5   *THERMOCHEMICAL DATA FOR*

C_xH_y	Name	C/H Mass (approx)	RMM	m_s	$(F/A)_s$
	PARAFFINS (Alkanes)				
CH_4	Methane	3	16.042	2	0.05817
C_2H_6	Ethane	4	30.068	3.5	0.06231
C_3H_8	Propane	4.5	44.094	5	0.06396
C_4H_{10}	Butane	4.8	58.120	6.5	0.06485
C_5H_{12}	Pentane	5	72.146	8	0.06541
C_6H_{14}	Hexane	5.14	86.172	9.5	0.06579
C_7H_{16}	Heptane	5.25	100.198	11	0.06606
C_8H_{18}	Octane	5.33	114.224	12.5	0.06627
C_9H_{20}	Nonane	5.4	128.250	14	0.06644
$C_{10}H_{22}$	Decane	5.45	142.276	15.5	0.06657
$C_{11}H_{24}$	Undecane	5.5	156.302	17	0.06668
	NAPHTHENES (Cyclanes)				
C_3H_6	Cyclopropane	6	42.078	4.5	0.06782
C_4H_8	Cyclobutane	6	56.104	6	0.06782
C_5H_{10}	Cyclopentane	6	70.130	7.5	0.06782
C_6H_{12}	Cyclohexane	6	84.156	9	0.06782
C_7H_{14}	Cycloheptane	6	98.182	10.5	0.06782
	OLEFINS (Alkenes)				
C_2H_4	Ethylene (Ethene)	6	28.052	3	0.06782
C_3H_6	Propene	6	42.078	4.5	0.06782
C_4H_8	1-Butene	6	56.104	6	0.06782
C_5H_{10}	1-Pentene	6	70.130	7.5	0.06782
C_6H_{12}	1-Hexene	6	84.156	9	0.06782
C_7H_{14}	1-Heptene	6	98.182	10.5	0.06782
C_8H_{16}	1-Octene	6	112.208	12	0.06782
C_9H_{18}	1-Nonene	6	126.234	13.5	0.06782
$C_{10}H_{20}$	1-Decene	6	140.260	15	0.06782
$C_{11}H_{22}$	1-Undecene	6	154.286	16.5	0.06782
$C_{12}H_{24}$	1-Dodecene	6	168.312	18	0.06782
	ACETYLENES (Alkynes)				
C_2H_2	Acetylene (Ethyne)	12	26.036	2.5	0.07553
C_3H_4	Propyne	9	40.062	4	0.07264

ΔH_f^o kJ/mol	$-\Delta H_r^o$ MJ/mol	MJ/kg	Dissociated n_5/n_T	STOIC. $T_{p\,ad}$ K Non.diss.	Diss.
-74.8977	0.8029	50.0471	0.00643	2330	2247
-84.7241	1.4288	47.5187	0.00775	2383	2282
-103.9164	2.0454	46.3865	0.00814	2396	2289
-124.8169	2.6602	45.7713	0.00832	2402	2293
-146.5380	3.2743	45.3840	0.00842	2405	2295
-167.3045	3.8893	45.1338	0.00851	2408	2296
-187.9455	4.5044	44.9549		2410	
-208.5864	5.1195	44.8200	0.00849	2411	2299
-229.1854	5.7347	44.7149		2413	
-249.8264	6.3498	44.6302	0.00868	2414	2300
-274.6541	6.9607	44.5339		2413	
			0.01116		
			0.01065		
-77.2883	3.1015	44.2255	0.00940	2439	2310
-123.2175	3.6914	43.8634	0.00899	2423	2305
52.3183	1.3238	47.1926	0.01270	2568	2420
20.4274	1.9277	45.8130	0.01116	2508	2362
1.1723	2.5442	45.3485	0.01065	2480	2348
-20.9340	3.1579	45.0291	0.00940	2471	2339
-41.7005	3.7729	44.8320	0.00899	2465	2333
-62.1740	4.3882	44.6943	0.00992	2459	2329
-82.9824	5.0031	44.5880	0.00980	2455	2326
-103.5814	5.6183	44.5070		2451	
-124.2224	6.2334	44.4419	0.00964	2448	2322
-144.8633	6.8485	44.3886		2446	
-165.4623	7.4637	44.3445	0.00953	2444	2319
226.8994	1.2564	48.2578	0.02477	2909	2583
185.5548	1.8509	46.1999	0.01704	2700	2476

C_xH_y	NAME	C/H Mass (approx)	RMM	m_s	$(F/A)_s$
C_4H_6	1-Butyne	8	54.088	5.5	0.07132
C_5H_8	1-Pentyne	7.5	68.114	7	0.07057
C_6H_{10}	1-Hexyne	7.2	82.140	8.5	0.07009
C_7H_{12}	1-Heptyne	7	96.166	10	0.06955
C_8H_{14}	1-Octyne	6.86	110.192	11.5	0.06949
C_9H_{16}	1-Nonyne	6.75	124.218	13	0.06930
$C_{10}H_{18}$	1-Decyne	6.67	138.244	14.5	0.06915
	AROMATICS				
C_6H_6	Benzene	12	78.108	7.5	0.07553
C_7H_8	Toluene	10.5	92.134	9	0.07425
C_8H_{10}	Xylene (average)	9.6	106.160	10.5	0.07333
$C_{10}H_8$	Naphthalene	15	128.164	12	0.07746
	ALIPHATIC ALCOHOLS				
CH_3OH	Methanol	3	32.042	1.5	0.15493
C_2H_5OH	Ethanol	4	46.068	3	0.11137
C_3H_7OH	Propanol	4.5	60.094	4.5	0.09685
	MISCELLANEOUS				
$H_2(g)$	Hydrogen (gas)	0	2.016	0.5	0.02924
$C(gr)$	Carbon (graphite)	∞	12.010	1	0.08710
CO	Carbon monoxide	∞	28.011	0.5	0.40640

ΔH_f^o kJ/mol	$-\Delta H_r^o$ MJ/mol	MJ/kg	Dissociated n_5/n_T	Stoic. $T_{p\ ad}$ K Non-diss.	Diss.
166.2160	2.4673	45.6161	0.01466	2611	2432
144.4446	3.0813	45.2371	0.01328	2573	2406
123.7199	3.6963	45.0002	0.01271	2552	2394
103.0790	4.3114	44.8333		2533	
82.4800	4.9266	44.7093	0.01156	2518	2366
61.8390	5.5417	44.6129		2507	
41.2400	6.1569	44.5364	0.01102	2498	2354
82.9824	3.1716	40.6054	0.01233	2529	2366
50.0323	3.7744	40.9666	0.01157	2504	2348
18.0591	4.3780	41.2757	0.01104	2487	2342
150.9341	5.0566	39.4545	0.01261	2534	2367
-201.3013	0.6765	21.1114	0.00725	2335	2243
-235.4656	1.2781	27.7427	0.00762	2356	2258
-235.1066	1.8932	31.5036	0.00804	2377	2273
0	0.2420	120.0338	0.01131	2534	2444
0	0.3938	32.7873	0.01120	2458	2309
-110.529	0.2830	10.1029	0.02124	2663	2399

REFERENCES AND BIBLIOGRAPHY

1. Goodger, E. M., *Hydrocarbon Fuels*, Macmillan, London and Basingstoke, 1975.

2. BS 1756: Part 2: 1971 Methods for the sampling and analysis of flue gases.

3. Mayhew, Y. R., and Rogers, G. F. C., *Thermodynamic and Transport Properties of Fluids*, 2nd ed., Blackwell, Oxford, 1973.

4. Stull, D. R., and Prophet, H., *JANAF Thermochemical Tables*, National Bureau of Standards, Washington, D.C., June 1971.

5. Penner, S. S., *Thermodynamics for Scientists and Engineers*, Addison-Wesley, London, 1968. (*Note* Includes earlier NBS values of K and $H_T - H_0$)

6. Goodger, E. M., *Principles of Engineering Thermodynamics*, Macmillan, London and Basingstoke, 1974.

7. Weast, R. C. (ed.), *Handbook of Chemistry and Physics*, 55th ed., The Chemical Rubber Co., Cleveland, Ohio, 1974/5.

8. --- 'Calorific Value of Liquid Fuel, IP 12/63T', *I.P. Standards for Petroleum and its Products*, Institute of Petroleum, London, 1974.

9. Spiers, H. M. (ed.), *Technical Data on Fuel*, 6th ed., British National Committee, World Power Conference, London, 1962. (*Note* Values tabulated are $H_T - H_{273}$)

10. --- *NBS Tables of Selected Values of Chemical Thermodynamic Properties*, Series 1, vol. 1, March 1947 to June 1949.

11. Hilsenrath, J., *et al.*, *Tables of Thermodynamic and Transport Properties of Air, Argon, Carbon Dioxide, Carbon Monoxide, Hydrogen, Nitrogen, Oxygen and Steam*, Pergamon, Oxford, 1960.

12. Spencer, H. M., 'Empirical Heat Capacity Equations for Various Gases', *J. Am. chem. Soc.*, 67 (1945) pp. 1859-60.

13. Goodger, E. M., 'Calculated Adiabatic Combustion Temperatures of Hydrocarbon-Air Mixtures', *Report No. SME 6*, Cranfield Institute of Technology, Bedford, June, 1974.

14. Fielding, D., and Topps, J. E. C., 'Thermodynamic Data for the Calculation of Gas Turbine Performance', *R & M No. 3099*, Ministry of Supply, H.M.S.O., 1959.

Boxer, G., *Engineering Thermodynamics, theory, worked examples and problems*, Macmillan, London and Basingstoke, 1976.

Eastop, T. D., and McConkey, A., *Applied Thermodynamics for Engineering Technologists*, Longman, London, 1970.

Harker, J. H., and Allen, D. A., *Fuel Science*, Oliver & Boyd, Edinburgh, 1972.

Rogers, G. F. C., and Mayhew, Y. R., *Engineering Thermodynamics, Work and Heat Transfer*, Longmans, London, 1967.

INDEX